HUMAN ERROR REDUCTION *and* SAFETY MANAGEMENT

HUMAN ERROR REDUCTION and SAFETY MANAGEMENT

Third Edition

Dan Petersen

VAN NOSTRAND REINHOLD
I(T)P® A Division of International Thomson Publishing Inc.

New York • Albany • Bonn • Boston • Detroit • London • Madrid • Melbourne
Mexico City • Paris • San Francisco • Singapore • Tokyo • Toronto

Copyright © 1996 by Van Nostrand Reinhold

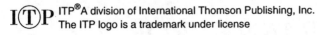 ITP®A division of International Thomson Publishing, Inc.
The ITP logo is a trademark under license

Printed in the United States of America

For more information, contact:

Van Nostrand Reinhold
115 Fifth Avenue
New York, NY 10003

Chapman & Hall
2–6 Boundary Row
London
SE1 8HN
United Kingdom

Thomas Nelson Australia
102 Dodds Street
South Melbourne, 3205
Victoria, Australia

Nelson Canada
1120 Birchmount Road
Scarborough, Ontario
Canada M1K 5G4

Chapman & Hall GmbH
Pappelallee 3
69469 Weinheim
Germany

International Thomson Publishing Asia
221 Henderson Road #05–10
Henderson Building
Singapore 0315

International Thomson Publishing Japan
Hirakawacho Kyowa Building, 3F
2-2-1 Hirakawacho
Chiyoda-ku, 102 Tokyo
Japan

International Thomson Editores
Campos Eliseos 385, Piso 7
Col. Polanco
11560 Mexico D. F. Mexico

All rights reserved. No part of this work covered by the copyright hereon may be reproduced or used in any form or by any means—graphic, electronic, or mechanical, including photocopying, recording, taping, or information storage and retrieval systems—without the written permission of the publisher.

1 2 3 4 5 6 7 8 9 10 BBR 01 00 99 98 97 96

Library of Congress Cataloging-in-Publication Data

Petersen, Daniel.
Human error reduction and safety management / Daniel Petersen. — 3rd ed.
p. cm.
Includes bibliographical references and index.
ISBN 0-442-02183-6 (hardcover)
1. Industrial safety. 2. Accidents—Prevention. I. Title.
T55.P35 1996
658—dc20
95-46268
CIP

Contents

Preface to the Third Edition ix
Preface to the First Edition xi
Introduction xiii

Part I. Human Error

Chapter 1. Human Error—What It Is 3

Chapter 2. Human Error—What It Does 9
 Examples 9
 Reasons to Look at Error 10
 Categorizing Errors 12

Chapter 3. Human Error—Causes 25
 Influences 25
 The Human-Error Causation Model 29

Part II. System-Caused Human Error

Chapter 4. Management Causes 37
 Management Safety Systems 42
 Assessment Criteria 56
 Logic Diagrams on Error Causes 59

Chapter 5. Culture Causes 63

Culture Definition 65
Culture and Safety 66
What Sets the Culture 67
Catastrophe Causes 67
Assessing Culture 73

Chapter 6. Design Causes 81
How Design Causes Error 85
How Design Causes Cumulative Trauma Disorders 91

Chapter 7. Reducing System-Caused Error 95
Reducing Management-Caused Error 95
Reducing Culture-Caused Error 96
Reducing Design-Caused Error 97
Ergonomic Systems Analysis 99
Task Analysis 100
CTD Analysis 105

Part III. Overload

Chapter 8. Causes and Outcomes of Overload 109
Overtime, Shift Work, and Other Trends 109
Reactions to Overload 112
Psychosocial Factors on the Job 117

Chapter 9. Capacity 125
Human Subsystems 125
Total Human Subsystem Considerations 131
Dealing with Capacity 136

Chapter 10. Load 145
Short-Term Load 145
Long-Term Load 157
Dealing with Load 160

Chapter 11. State 163
Motivational State 163
Attitudinal State 174
Arousal State 174
Biorhythmic State 176
What to Do about State 177

Chapter 12. Reducing Overload-Caused Errors 179
Strategies for Changing the Work Environment 181
Strategies for Reducing Overload 182

Part IV. Decision to Err

Chapter 13. Logical Decision to Err   195
- Reinforcement as a Determinant   196
- The Peer Group   199
- Attitudes   206
- The Boss's Style, Measures, and Priorities   209
- What to Do to Get Results   211

Chapter 14. Proneness   213
- What It Is   213
- What to Do about It   218

Chapter 15. Perception of Risk   223
- Common Sense and Perception   227
- Attitudes   230
- Managing Risk   232
- Force-Field Analysis   237
- Achieving Change in People   242

Chapter 16. Reducing Decisions to Err   247
- Setting Expectations   247
- Positive Reinforcement   247

Part V. Roles

Chapter 17. Human Error Reduction Concepts   255
- What Makes It Happen   256
- Traditional Controls   257
- Nontraditional Controls   260

Chapter 18. Line Management Roles   273
- Upper Management   273
- Change   278
- Upper-Management Changes   281
- Lower-Management   290
- Lower-Management Tasks   294
- The Worker Role   302

Chapter 19. Role of Staff   305
- The Change Agent and Approaches to Change   310

Part VI. Appendixes

Appendix A. Stress Tests (Individuals)   323
Appendix B. CTD Analysis   335

Appendix C. Ergonomic Data 339
Appendix D. Behavioral Research 355
Appendix E. Major Incidents 385

Index 393

Preface to the Third Edition

What has happened since publication of the first edition (1982) of this book is somewhat unreal to most of us in the safety profession. We have truly faced a paradigm shift, which is touched on in the book. None of us can ignore the shift.

We are in a different world—and we will be there from now on. Much of what was in the first edition is now an integral part of that new world: the section on ergonomics, for example, as well as the references to psychological stress in the overload section, and so forth.

After a paradigm shift one must rethink everything. This edition is intended to assist its readers in that rethinking.

Preface to the First Edition

Most management problems, and, I believe almost all safety problems are in fact people problems. I attempt to examine some of the reasons behind these people problems, particularly the reasons behind the behavior of the error-making worker. This book is based upon the belief that most accidents are caused by people doing things unsafely even though they know better, a belief that says people choose logically to act unsafely and that the choice is one that most people would make in the same situation.

Human Error Reduction and Safety Management is not only on accident prevention; it is also concerned with error reduction, and is much broader in scope than just safety.

Many people have assisted me in the preparation of this book. I have quoted from many in the following pages. I wish to thank each for allowing me to use his or her ideas and words. I would also like to express particular thanks to Dr. Russ Ferrell of the University of Arizona, whose thoughts started me in this direction; to D. A. Weaver, who has always helped my thinking tremendously in this profession; and to Drs. Idahlynn Karre, Kenneth Douglas, and Donald Holley, all of the University of Northern Colorado for their help and review. Finally, I'd like to thank Frank E. Bird, Jr., of the International Loss Control Institute, whose early review of the concepts of this book was extremely helpful.

Introduction

Safety practitioners have, since about 1931, held a fundamental belief that the single largest reason behind accidents is people. This belief was first expressed in writing in 1931 by H. W. Heinrich in his early text *Industrial Accident Prevention.* Prior to this time there had been safety work, but no real theory behind it—no thought-out concepts to guide the profession. In 1931 Heinrich provided this by developing some theory for the profession and some concepts for practitioners to follow.

Before 1931 safety practitioners tended to concentrate on physical preventive measures such as machine guarding, housekeeping, and inspection, apparently in the belief that physical conditions cause accidents. Little preventive work was done with people, other than some awareness programs (with posters, etc.). Heinrich changed all of this with his new and different thoughts about how accidents happen and about what needs to be done for their control. Heinrich's theory and concepts were summed up in ten statements of fact that he called axioms of industrial safety.

Today these statements are not particularly exciting; in fact, they seem perhaps a little out of date. But in 1931 they were revolutionary from several standpoints. First, they explained the accident process for the first time. In axiom 1, Heinrich spelled out a model of accident causation. Further, he graphically illustrated this model with dominoes. With this model of causation, safety professionals had for the first time the concept that an accident is the result of one or both of two things: an act and a condition. He introduced the person into the causation sequence and suggested that accident control is a two-pronged problem: a "thing problem" and a "people problem." Thus the concept examined in this book, *Human Error Reduction and Safety Management,* had its birth in 1931 with Heinrich.

Heinrich went even further in introducing the concept that accidents are caused by people. He stated simply that the majority of accidents are caused by the unsafe acts of people, rather than by physical conditions. While this simple thought seems obvious today, it did not seem obvious in 1931. Prior to 1931, almost the entire emphasis of control was on conditions. Heinrich's work showed safety professionals that they were doing the wrong things. Heinrich backed up his concept with this statistic: 88% of all accidents are caused by the unsafe acts of people rather than by unsafe conditions. He stated that not only were safety practitioners doing the wrong things when working on conditions, but that 88% of their time was spent in doing the wrong things altogether.

We know today that his figures were meaningless. They were, in fact, based upon a research methodology that would invalidate them totally. But we also know today that his concept was most meaningful and extremely valid. People *are* the primary causes of accidents.

The safety movement followed Heinrich's thinking. Safety professionals tended to believe that people cause accidents, and they did decide to spend a portion of their time on the human side of accident control. Their primary emphasis was still on physical conditions, but they at least began to recognize the importance of the unsafe act.

As we examine Heinrich's axioms, which have guided the safety profession for years, we find a concept of people as the fundamental reason behind accidents, and an almost totally ridiculous approach to the control of human behavior. The axioms include the single most valuable clue to accident control ever found, human behavior, and some approaches to that control that were mostly doomed to fail.

Heinrich's philosophy has almost totally directed the profession since 1931. By looking at five of the most common areas in our safety programs today, we can quickly see how our programs are built on his principles (and how they are beginning to change).

CURRENT SAFETY PROGRAMS

Accident Investigations and Inspections

Accident investigations and inspections are integral to all safety programs. The Occupational Safety and Health Act (OSHA) depends almost entirely on them. These are based entirely upon Heinrich's axiom 1, the domino theory of accident causation.

Newer safety theory disputes this domino theory and replaces it with a multiple-causation theory. The multiple-causation theory states that accidents are caused by the combination of a number of things, all wrong, which combine at one point in time and result in an injury. This theory suggests that the act, the condition, and the accident itself are all symptoms of something wrong in the management system. The role of safety is not to remove the symptom, but to find out what is wrong with the system.

Employee Training

A second safety area is the area of employee training. Safety programs invariably include aspects such as safety orientation, employee training, and safety talks for first-line supervisors. These are based upon the Heinrich axiom that states: "The unsafe acts of persons are responsible for a majority of accidents." The underlying assumption of this axiom is that the employees do not know the difference between right and wrong or safe and unsafe procedures. We know that employees do know, in many cases, what is safe behavior and what is not. In such cases, training is not the solution to the problem.

Supervisory Training

A third area is supervisory training. Safety programs are almost always heavy in the area of supervisory training. This training is based on the Heinrich axiom that says: "The supervisor or foreman is the key man in accident prevention." One of the underlying assumptions here is that merely identifying the supervisor as the key person magically makes the supervisor do something about safety. He or she will not.

Records

Fourth, safety programs include record keeping and the analysis of these records. The belief is that if we find the most common kinds of incidents and attack them, automatically our costs will come down. This thinking is based on the Heinrich axiom that states: "The severity of an injury is largely fortuitous." We have interpreted this to mean that the causes of accident frequency and severity are the same. Today we know that this is wrong. Severity is predictable in certain situations and under certain circumstances.

Safety Media

Fifth, safety programs include the use of all kinds of safety media, such as posters, banners, films, and literature. This is based on the Heinrich axiom that says that one of "the four basic motives or reasons for the occurrence of unsafe acts [is] . . . improper attitude."

Do posters, banners, and the like do any good? We really do not know. To be more accurate, so far we know just a little bit about the answer to this question. Research tells us that in some cases these kinds of safety media help, in some cases they do harm, and in some cases they have absolutely no effect. The effect apparently depends more on the employee's perception of management's credibility than on the poster itself.

In short, our current safety programs are primarily based on Heinrich's 1931 thinking, plus some rather questionable assumptions about people and how we can change their behavior.

THE STATE OF THE ART

The state of the art in safety management is somewhat confused. For the most part, we have safety programs in operation in most industries that are based on the theories of Heinrich, and we have safety professionals who no longer really believe in these theories. It is clear, however, that two of Heinrich's concepts are still respected today. One is the belief that the primary cause of accidents is people.

We believe it more than ever today, and we have found ways to modify human behavior effectively. Even a federal law (OSHA) that denied that human behavior causes accidents did not change this fundamental belief.

The other concept originally espoused by Heinrich and still believed in today is that safety is a management problem: a problem that can be solved by management.

That these two fundamental insights are correct is clearer today than ever before. Our knowledge of them has not helped us to achieve results, partly because Heinrich misdirected us after giving us the truths and partly because we simply did not know how to control the behavior of people with the management beliefs, styles, and structures that have existed.

Our thinking has been influenced by other disciplines and other approaches in recent years, which should contribute to future directions and successes. Two disciplines that seem to be particularly important are human factors engineering and systems engineering. Most readers are quite familiar with both of these disciplines. As the book progresses, their influence on our subject will be obvious. A third major area of influence is management theory, particularly management theory in which new concepts from the behavioral sciences are used.

In this book, we will elaborate further upon Heinrich's two key insights:

1. People are the prime cause of most accidents, and safety is primarily a

human problem. Our future successes will come from learning more about how to control human behavior.
2. The control of accidents is a management problem. While management must control both the conditions that exist and the behavior of employees, the area of primary concern for safety professionals is behavioral control.

PART I | *Human Error*

Chapter *1*

Human Error—What It Is

In this book we offer a model of accident causation. The model implies that human error is a basic cause behind all accidents and incidents. There may also be systems failure; but whether or not there is a systems failure, there certainly will be a human error. Since the remainder of this book is about human error, we start with some definitions.

George Peters defines human error in these terms:[1]

> In theory, we would want to use a broadly oriented definition which states that a human error consists of *any significant deviation from a previously established, required or expected standard of human performance.*
>
> In practice, the term may have any one of several specific meanings depending upon the nature of contractual agreements, the unique requirements of a particular program, the customary error classification procedures, and the emotional connotations involved with the use of a term which might be incorrectly perceived as possibly placing the blame on individuals or their immediate supervision.
>
> In the reality of situations where arguments of precisely what is or is not a human error are of less importance than what can be done to prevent them, the operational definition may be restricted to those errors (a) which occur within a particular set of activities, (b) which are of some significance or criticality to the primary operation under consideration, (c) [which] involve a human action of commission or omission, and (d) about which

---

1. Reprinted with permission from G. Peters, Human error: analysis and control, *Journal of the ASSE,* January 1966.

there is some feasible course of action which can be taken to correct or prevent their reoccurrence.

Alphonse Chapanis begins one of his papers with the following case history:[2]

> In March 1962 a shocked nation read that six infants died in the maternity ward of the Binghamton, New York, General Hospital because they had been fed formulas prepared with salt instead of sugar. The error was traced to a practical nurse who had inadvertently filled a sugar container with salt from one of two identical, shiny, 20-gallon containers standing side by side, under a low shelf, in dim light, in the hospital's main kitchen. A small paper tag pasted to the lid of one container bore the word "Sugar" in plain handwriting. The tag on the other lid was torn, but one could make out the letters "S..lt" on the fragments that remained. As one hospital board member put it, "Maybe that girl did mistake salt for sugar, but if so we set her up for it just as surely as if we'd set a trap." . . . When a system fails it does not fail for any one reason. It usually fails because *the kinds of people* who are trying to operate the system, with *the amount of training* they have had, are not able to cope with *the way the system is designed,* following procedures they are supposed to follow, *in the environment* in which the system has to operate.

These definitions of human error begin to bring out what is, perhaps, the key concept in thinking about human error reduction; that human beings do not commit errors because they are dumb or because they are wrong. Human beings commit errors because it is logical that they do so in the situations they are in. Human errors are caused; they do not just happen. To repeat, human errors are caused by the situations in which people find themselves—physical situations, psychological situations, and so on. A worker may be in a particular situation at a particular moment that makes it totally normal and logical to commit an error that may result in an accident and an injury. In this book, we will look at the situations that cause human error and determine what, if anything, can be done by management to reduce the frequency of these situations.

While the book concentrates on those errors that result in injury or other financial loss to an organization, we should keep in mind that error is a part of our everyday living; the consequences of error can be major or slight, but to err is normal—we all do it. Human history has been error-plagued. The following paragraphs give a few examples.

---

2. Reprinted with permission from A. Chapanis, in W. Johnson, *New Approaches to Safety in Industry,* London, InComTec, 1972.

Human Error—What It Is

Christopher Columbus, a popular subject at the time of this writing, provided us a story filled with error, as described in a recent publication, *The Blunder Book*:[3]

> First, he thought he had indeed landed in India and dubbed his discovery West Indies and the natives Indians. Later he discovered what is now Costa Rica and gave it that name, which means "rich coast," because he saw natives with gold necklaces and believed the land was rich with gold and silver. However, Costa Rica has proved to be among the Latin American countries with the *least* amount of mineral wealth. But as with the West Indies, the name Costa Rica stuck, remaining on the map as another example of Columbus's errors.
>
> For all his naming of places in the New World, Columbus missed out not only on naming the most important land of all—the continents themselves—but on having them named in his honor. That honor went to a little-known Italian merchant—explorer Amerigo Vespucci, who claimed to have discovered the continent of America in 1497. Since Columbus at that time still had no idea he had reached the Western Hemisphere but continued to believe he was exploring parts of the Indies (he did not step onto the mainland of South America until 1498), he did not dispute Vespucci's claim. Vespucci went on to publish letters about his discovery, and in 1507, when a mapmaker needed a name for the new continent (which was South America), he suggested America "because Amerigo discovered it."

The Blunder Book also cites this classic case:[4]

> In 1685 Charles II fell ill of kidney disease and began to experience convulsions. For the next five days a parade of doctors—as many as twelve during the day and six at night—tried out their ideas on his progressively deteriorating body.
>
> First they began by bleeding him, that favorite treatment of physicians of the time. They put cupping glasses on his shoulders, made incisions in his skin, and drew blood from his veins. Then they cut off his hair, put blistering agents on his scalp, and applied plasters of pitch and pigeon dung to the bottoms of his feet. After this they blew the offensive-smelling herb hellebore up his nostrils so that he would sneeze and thereby release the humors from his brain. After that they poured antimony and sulfate of zinc down his throat so that he would throw up and cleanse his insides. To clean out his bowels, they administered purgatives. And to stop the convulsions, they gave him spirit of human skull.

---

3. Reprinted from M. Goldberg, *The Blunder Book*, New York, William Morrow & Co., 1984.
4. Ibid.

All the while this was going on, they periodically administered juleps for spasms, a gargle for sore throat, liquids to quench his thirst, tonics for his heart, and beer and broth for food. Sedatives, laxatives, herbs like cowslip and mint, even bezoar (a concoction found chiefly in the alimentary organs of cud-chewing animals and believed to be an antidote for poison)—all were given to the royal patient. As further precaution he was not allowed to sleep or talk. Not only physicians but priests, ministers, and servants entered the royal chamber to be of assistance. Finally, on the fifth day, the doctors drew twelve more ounces of the king's blood and gave him more heart tonics. This seemed to do the trick. By noon Charles II was dead.

In summing up the monarch's royal treatment at the hands of his physicians, Thomas Babington Macaulay, the nineteenth-century English historian and statesman, declared that the doctors had tortured the king like an Indian at the stake.

This case was printed in an old newspaper under the heading of the "Great Air Mail Fiasco":

Air service was officially introduced by the U.S. Post Office Department in 1918. It could just as easily been introduced by "Wrong Way" Corrigan with an assist from the Marx Brothers. Nothing went right.

On the morning of May 15, 1918, crowds gathered at a large polo field in Washington's Potomac Park to witness the departure of the first air mail flight in the world. President Woodrow Wilson was on hand along with other important government officials. At 11 A.M., pilot George Boyle settled into the cockpit of his 150 horsepower Curtiss "Jenny," ready to go. But the engines would not start. Mechanics sweated over the aircraft for nearly 30 minutes, while the President grew visibly irritated. Finally somebody checked the fuel tanks—they were empty.

His tanks filled, Boyle took off. But instead of flying north to Philadelphia, his first scheduled stop, he followed the wrong set of railroad tracks and flew southeast. Intending to ask directions, he touched down near Waldorf, Maryland, 20 miles from Washington, D.C. But he wrecked the propellor on landing, and couldn't get off the ground again. His cargo—140 pounds of air mail—had to be delivered by train to Philadelphia, and by plane to New York.

The Postal department also discovered an error in the first issue of air mail stamps. Illustrated with an engraving of the Curtiss "Jenny," they had been printed upside down.

There was the true story of a 50-year-old man from Vienna, Austria who entered a local hospital to be treated for rheumatism. While settling into the hospital, he slipped on a wet floor, breaking his leg. Because of an error in the hospital systems,

Human Error—What It Is

one morning he was brought into the operating room, where a pacemaker was inserted in his chest.

Some readers may remember the 1948 presidential election, and the papers the next morning showing Harry Truman holding up a Chicago newspaper announcing "Dewey Defeats Truman."

Many readers will remember Ford's "Edsel," and some may have visited Italy and viewed the Leaning Tower of Pisa.

Here are some great examples from sports:[5]

> *Babe Ruth:* An awesome batter, with 714 home runs, he held the record for most career homers for thirty-nine years . . . but at one time he also struck out more times than any other player in baseball history (1,330 strikeouts). And on September 11, 1931, in a game with the Chicago White Sox, he grounded into a triple play.
>
> *Ty Cobb:* A fierce competitor, batting champ, and once the holder of most base steals in a season and in a career, until 1982 Cobb also held the record for being thrown out the most times attempting to steal in a season (38 times in 1915).
>
> *Cy Young:* With 511 victories to his credit, Young holds the record for most career wins by a pitcher. However, he also had 313 losses, the record for most losses by a pitcher (he lost 20 or more games in a season three times and once had a 13–21 season).
>
> *Hank Aaron:* Slugger of most home runs in a career with 755, Aaron also holds the career record for hitting into the most double plays.
>
> *Walter Johnson:* One of the greatest pitchers of all time (until recently the holder of most strikeouts over a career—3,508), Johnson holds the record for hitting batters (204) and is tied for most wild pitches in a career (156).
>
> *Jimmy Foxx:* The great right-handed batter (he once hit 58 home runs in a season) holds the record for leading a league in strikeouts for the most consecutive seasons (7).
>
> *Roberto Clemente:* The Pittsburgh Pirate batting star once struck out 4 times in an All-Star game—a record.
>
> *Sandy Koufax:* The pitching ace for the Brooklyn Dodgers, with four no-hitters to his credit, had a difficult time as a batter. He holds the record for striking out the most consecutive times at bat: 12.
>
> *Joe DiMaggio:* The Yankee Clipper once hit into 7 double plays during a World Series—still a record.

The bottom line is that human error happens to all of us: "to err is human." This has been true throughout history. None of us is immune to human error. We all err, and do so many times every day.

---

5. Reprinted from M. Goldberg, *The Blunder Book*, New York, William Morrow & Co., 1984.

REFERENCES

Chapanis, A. Quoted in W. Johnson, *New Approaches to Safety in Industry.* London: InComTec, 1972.
Goldberg, M. *The Blunder Book.* New York: William Morrow & Co., 1984.
Peters, G. Human error: analysis and control. *Journal of the ASSE,* January 1966.

Chapter *2*

Human Error—What It Does

As indicated in the last chapter, human error is normal. We all make mistakes—every day. Clearly we cannot control all human error. But the more we understand it, the more we can begin to effect some control. Part I of this book describes the phenomenon of human error—what it is, what it does, and, in Chapter 3, what causes it. Parts II, III, and IV examine categories of the causes of human error, and suggest what we can do about it in somewhat specific terms.

Chapter 1 provided a definition of human error: deviation from a standard of human performance, according to George Peters. Throughout the human error literature a theme emerges that errors are usually caused by the system; or the environment that we in management have created, and therefore can alter.

This chapter considers the results of human error—what it does; and, as indicated earlier, the results can range from the minimal to the catastrophic.

EXAMPLES

First, here are some personal examples:

- I have often screwed up the balance in our family checkbook by writing a check for several hundred dollars and forgetting to record it—or have done worse by recording it as a deposit, doubling the error.
- To me, a "system-caused error" is provided too easily by my bank and its automatic teller. Since I have a bank card and the "key number," and my wife carries the checkbook, my withdrawals seldom get recorded.

- One year my family spent the better part of the summer at a small cabin we owned in Estes Park, Colorado, and then, when fall came, we headed home to Tucson, Arizona. It was only when we got home, 1,000 miles away, that I realized I had left my briefcase, checkbook, work schedule, and everything I needed to earn a living inside the locked-up, winterized cabin.
- I recall the day I got up at 6:00 A.M. to make an 8:00 A.M. flight from Tucson to the East, got dressed, and had my son drive me to the airport and drop me off at the entrance only to find that I was there on the wrong day.
- Another time I flew into Philadelphia, took a cab to the Airport Hilton, and, after making a scene about their losing my reservation, rechecked to find that I should have gone to the Airport Holiday Inn.
- I "routinely" push the wrong button on my telephone recording machine, thereby losing a week's worth of important messages.

Each reader surely can give similar everyday examples of human error.

Looking beyond personal examples, consider these types of errors:

- *Design errors,* as in the Pinto gas tank, where in an apparent attempt to save a few dollars in production, human lives were lost; or the DC-10 aircraft, where jet engines fell off in the air and lives were lost.
- *Communication errors,* as in the crash of two 747s on an island in the Atlantic, producing the worst loss of life in an air crash in history.
- *Management system errors,* as with Bhopal, Chernobyl, the space shuttle, and others, among the serious ones, not to mention the minor ones already cited.

Sometimes, we are fascinated by, and pay more attention to, the catastrophic losses. While their causes are usually the same as those of the minor losses, at least they get our attention. As a matter of fact, they get everybody's attention. Consider the publicity accorded the *Exxon Valdez* incident, which made the headlines every day for six months (though no single human was ever injured). Or the Three-Mile Island incident, when all the fail-safe systems worked, and it still almost destroyed the organization.

REASONS TO LOOK AT ERROR

A catastrophic incident, whether or not it results in serious injury or loss of life, has serious corporate implications:

- Extremely poor publicity, with attendant financial losses.
- Probably new legislation with good intent, but seldom aimed at the incident causes.

Human Error—What It Does

- Much executive involvement combating the above.
- Possible corporate demise.
- Reduced employee trust and morale.
- Lawsuits.
- Massive loss of money.
- Executives possibly going to jail.

All of the above, and much more, results from incidents due to human error, whether system-caused or not. A few of these items need elaboration.

Executives in Jail

One change we are experiencing in the external safety/legal environment is that government (federal, state, and local) is now taking an active role in criminal prosecution for occupational health and safety problems. At the time of this writing the trend has just begun, but it is distinctly with us and spreading. For instance, there have been referrals from OSHA to the Justice Department for criminal prosecution. Many of the accused have been convicted and are serving jail terms. State and local governments are doing the same thing. Los Angeles County alone has indicted many executives, and convicted some who served jail terms. A Los Angeles assistant district attorney is now traveling the country assisting other district attorneys' offices in setting up occupational health and safety sections. She states that all accidents are to be viewed as crimes.

Breakdown of the Worker's Compensation System

If you think you or your company is protected under the worker's compensation system from lawsuits, think again. The system is falling apart. Lawsuits are again rampant, and this time you cannot count on the old common law defenses. Chances are very good that you will pay.

Any organization needs to deal right now with what is causing human error, which causes accidents. The problem is real. Injuries are increasing in the United States at an alarming rate (as much as 18% per year). Costs are doubling every two years. If these two figures are put into a computer, it suggests that we could be heading for trouble. If your company this year experiences 25 lost time injuries at the average cost of $40,000, the table shows where you are heading.

	No. injuries	Cost/LT	Loss
This year	25	$ 40,000	$1,000,000
Next year	30	60,000	1,800,000
Year after	35	80,000	2,800,000
Year after	41	120,000	4,900,000
Year after	48	169,000	7,700,000

You also have the option of controlling these losses and saving your company $18,200,000 over the next five years (direct costs only). You can do this only through human error reduction.

Catastrophe Control

The above five-year amount can be lost in a few moments through a catastrophe. Often we think of catastrophes as "different"—due to different causes, controlled differently. This is untrue; the catastrophes that we have seen and remember tend to come from the same system causes as the rest of our problems.

Recently a study was made of some of our worst catastrophes to unearth the management and systems failures behind them. The study looked at the underlying causes of four major incidents: the Three Mile Island near meltdown, the meltdown at Chernobyl, the Challenger incident, and the Bhopal chemical leak. Investigators found surprisingly similar causes behind all four, and systems that caused human error in all four. This is discussed in detail in Chapter 5.

CATEGORIZING ERRORS

Error Outcomes

Nuberg provides an error spectrum (Figure 2-1), and Bond (Figure 2-2) suggests a rather detailed logic tree of possible outcomes of event sequences. Human error here is indicated by "no response" or "inappropriate" response.

When we talk about error outcomes, we are certainly discussing much more

ERROR SPECTRUM

Work free of mistake and error	Minor errors, mistakes and slight blemishes	Errors causing delay, seconds, rework, rejects, waste	Damage to Property, Material Loss, Process Delay	Errors causing injury	Acts of negligence and deliberate destruction Theft, Arson, Pollution.

Figure 2-1. Nuberg's error spectrum. (From A. Nuberg, Reducing damage accidents. In J. Widener, Ed., *Selected Readings in Safety*. Macon, Ga.: Academy Press, 1973.)

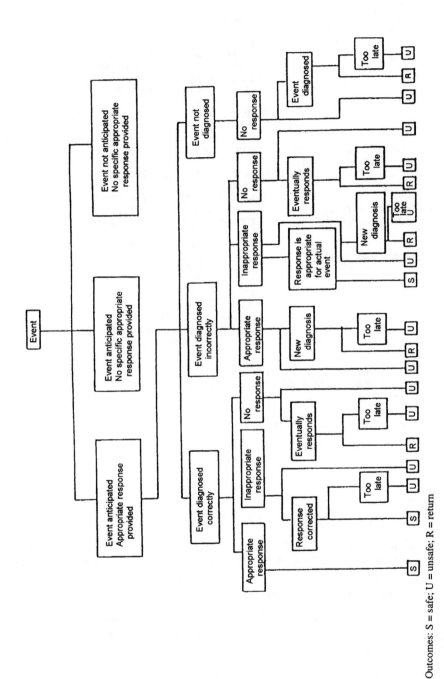

Outcomes: S = safe; U = unsafe; R = return

Figure 2-2. Possible outcomes of event sequences. (From N. Bond, *Aviation Psychology*. Los Angeles: University of Southern California Press, 1968.)

Table 2-1. Typical human errors in equipment operation

Type of error	Possible causal factors
Failure to detect signal	Input overload
	Too many significant signals
	Too may separate input channels
	Input underload
	Too little variety of signals
	Too few signals
	Adverse noise conditions
	Poor contrast
	High intensity of distraction stimuli
Incorrect identification of signal	Code form or typology unclear
	Lack of differential cues
	Inappropriate filtering set (expectation)
	Conflicting cues
	Conflicting identification requirements
Incorrect value weighting or assignment priority	Nonlinear predictions required
	Multiple or complex value sealing required
	Values poorly defined or understood
	Contingencies vaguely defined
Error in action selection	Matching of actual and required patterns faulty
	Consequence of courses of action not understood
	Appropriate action not available
	Correct action inhibited
	Cost considerations
	Procedural prohibitions
Error of commission	Correct tool or control not available
	Action–control relationship not understood
	Action feedback unavailable or delayed

Source: Adapted from R. McFarland, Application of human factors engineering to safety engineering problems, in J. Widener, Ed., *Selected Readings in Safety.* Macon, Ga: Academy Press, 1973.

than accidents. Error outcomes can refer to accidents and incidents (safety problems) and also to quality problems, production problems, security problems, and others. While our primary focus in this book is on accidents, most of what is said will pertain directly to other outcomes of human error.

Error Types and Causes

A number of writers have discussed and attempted to classify types of human error. Ross McFarland's table of five categories (Table 2-1) also lists causal factors.

J. L. Recht has written as follows:[1]

> More elaborate methods are also available to determine by probability computations how critical specific human errors are for degradation of system

---

1. Reprinted from *National Safety News,* a National Safety Council publication, 444 N. Michigan Ave., Chicago, IL.

Human Error—What It Does

Table 2-2. Representative human-error rates for 15 task elements

Action	Object	Error	BER*
Observe	Chart	Inappropriate switch actuation	1128
Read	Gauge	Incorrectly read	5000
Read	Instructions	Procedural error	64500
Connect	Hose	Improperly connected	4700
Torque	Fluid lines	Incorrectly torqued	104
Tighten	Nuts, bolts	Not tightened	4800
Install	Nuts, bolts	Not installed	600
Install	O-ring	Improperly installed	66700
Solder	Connectors	Improper solder joint	6460
Assemble	Connector	Bent pins	1500
Assemble	Connectors	Omitted parts	1000
Close	Valve	Not closed properly	1800
Adjust	Mechanical linkage	Improper adjustment	16700
Install	Line orifice	Wrong size installed	5000
Machine	Valve port	Wrong size drilled and tapped	2083

Source: J. L. Recht. *Systems Safety Analysis.* Error Rates and Costs. Chicago: National Safety Council. 1963.
*Basic error rate (errors per million operations).

performance. These techniques, however, also depend upon basic human error rates.

Basic human error rates are usually expressed in terms of the number of errors per million operations—based upon prior experience in similar situations. Some representative error rates are shown (Table 2-2) to illustrate the range and magnitude of such measurements. (Warning: this data should not be used for computational purposes without additional background information—specifically, under what conditions these rates can be expected to be valid and the probable error in each rate.)

Unfortunately the greatest restriction on the use of quantitative human error techniques is the lack of sufficient error rate data.

When describing human error, perhaps as with any relatively new body of knowledge, each researcher or author tends to look at it differently, with different types defined, different classifications, and so on. For instance, Trevor Kletz, of the United Kingdom, describes these types of human error, giving examples:[2]

- *Errors due to slips or aberrations:* Example—Opening equipment which has been under pressure

A suspended catalyst was removed from a process stream in a pressure filter. When a batch had been filtered, the inlet valve was closed and the

---

2. Reprinted with permission from T. Kletz, *An Engineer's View of Human Error,* the Institution of Chemical Engineers, Rugby, England, 1985.

liquid in the filter blown out with steam. The steam supply was then isolated, the pressure blown off through the vent and the fall in pressure observed on a pressure gauge. The operator then opened the filter for cleaning. The filter door was held closed by eight radial bars which fitted into U-bolts on the filter body. To withdraw the radial bars from the U-bolts and open the door the operator had to turn a large wheel, fixed to the door. The door, with filter leaves attached, could then be withdrawn.

One day an operator, a conscientious man of great experience, started to open the door before blowing off the pressure. He was standing in front of it and was crushed between the door and part of the structure and was killed instantly.

- *Errors that could be prevented by better training or instructions*

The accident at Three Mile Island Nuclear power station in 1979 had many causes and many lessons can be drawn from it but some of the most important are concerned with the human factors. In particular the training the operators had received had not equipped them to deal with the events that occurred.

- *Errors due to a lack of physical ability:* Example—Someone is asked to do the physically difficult or impossible

A steel company found that overhead travelling magnet cranes were frequently damaging railway wagons. One of the causes was found to be the design of the crane cab. The driver had to lean over the side to see his load. He could then not reach one of the controllers, so he could not operate this control and watch the load at the same time.

- *Errors due to a person being asked to do the mentally difficult or impossible*

If a man is asked to detect a rare event he may fail to notice when it occurs, or may not believe that it is genuine. The danger is greatest when he has little else to do. It is very difficult for night watchmen, for example, to remain alert when nothing has been known to happen and when there is nothing to occupy their minds and keep them alert. (Compare *St Matthew,* 26, 40, "Could ye not watch with me one hour?")

There is another category that is perhaps somewhat different: *errors due to the perception of peer thought.* Several years ago I purchased and read a great management book by Jerry Harvey, *The Abilene Paradox.* In Dr. Harvey's key essay, "The Abilene Paradox: The Management of Agreement," he illustrates this category:[3]

---

3. Reprinted with the permission of Lexington Books, an imprint of The Free Press, a Division of Simon & Schuster, Inc., from *The Abilene Paradox and Other Meditations on Management* by Jerry B. Harvey. Copyright © 1988 by Jerry B. Harvey.

That July afternoon in Coleman, Texas (population 5,607), was particularly hot—104 degrees according to the Walgreen's Rexall's thermometer. In addition, the wind was blowing finegrained West Texas topsoil through the house. But the afternoon was still tolerable—even potentially enjoyable. A fan was stirring the air on the back porch; there was cold lemonade; and finally, there was entertainment. Dominoes. Perfect for the conditions. The game requires little more physical exertion than an occasional mumbled comment, "Shuffle 'em," and an unhurried movement of the arm to place the tiles in their appropriate positions on the table. All in all, it had the makings of an agreeable Sunday afternoon in Coleman. That is, until my father-in-law suddenly said, "Let's get in the car and go to Abilene and have dinner at the cafeteria."

I thought, "What, go to Abilene? Fifty-three miles? In this dust storm and heat? And in an unairconditioned 1958 Buick?"

But my wife chimed in with, "Sounds like a great idea. I'd like to go. How about you, Jerry?" Since my own preferences were obviously out of step with the rest, I replied, "Sounds good to me," and added, "I just hope your mother wants to go."

"Of course I want to go," said my mother-in-law. "I haven't been to Abilene in a long time."

So into the car and off to Abilene we went. My predictions were fulfilled. The heat was brutal. Perspiration had cemented a fine layer of dust to our skin by the time we arrived. The cafeteria's food could serve as a first-rate prop in an antacid commercial.

Some four hours and 106 miles later, we returned to Coleman, hot and exhausted. We silently sat in front of the fan for a long time. Then, to be sociable and to break the silence, I dishonestly said, "It was a great trip, wasn't it?"

No one spoke.

Finally, my mother-in-law said, with some irritation, "Well, to tell the truth, I really didn't enjoy it much and would rather have stayed here. I just went along because the three of you were so enthusiastic about going. I wouldn't have gone if you all hadn't pressured me into it."

I couldn't believe it. "What do you mean 'you all'?" I said. "Don't put me in the 'you all' group. I was delighted to be doing what we were doing. I didn't want to go. I only went to satisfy the rest of you. You're the culprits."

My wife looked shocked. "Don't call me a culprit. You and Daddy and Mama were the ones who wanted to go. I just went along to keep you happy. I would have had to be crazy to want to go out in heat like that."

Her father entered the conversation with one word: "Shee-it." He then expanded on what was already absolutely clear: "Listen, I never wanted to go to Abilene. I just thought you might be bored. You visit so seldom I wanted to be sure you enjoyed it. I would have preferred to play another game of dominoes and eat the leftovers in the icebox."

After the outburst of recrimination, we all sat back in silence. Here we were, four reasonably sensible people who—of our own volition—had just taken a 106-mile trip across a godforsaken desert in furnace-like heat and a dust storm to eat unpalatable food at a hole-in-the-wall cafeteria in Abilene, when none of us had really wanted to go. To be concise, we'd done just the opposite of what we wanted to do. The whole situation simply didn't make sense.

At least it didn't make sense at the time. But since that day in Coleman, I have observed, consulted with, and been a part of more than one organization that has been caught in the same situation. As a result, the organizations have either taken side trips or, occasionally, terminal "journeys to Abilene," when Dallas or Houston or Tokyo was where they really wanted to go. And for most of those organizations, the negative consequences of such trips, measured in terms of both human misery and economic loss, have been much greater than for our little Abilene group.

I now call the tendency for groups to embark on excursions that no group member wants "the Abilene Paradox." Stated simply, when organizations blunder into the Abilene Paradox, they take actions in contradiction to what they really want to do and therefore defeat the very purposes they are trying to achieve. Business theorists typically believe that managing conflict is one of the greatest challenges faced by any organization, but a corollary of the Abilene Paradox states that the inability to manage *agreement* may be the major source of organization dysfunction.

Obviously, there are many reasons behind human error, both personal and social. The Abilene Paradox can be found to apply to many situations. Consider the following:

- Why did the Kennedy cabinet go into the "Bay of Pigs" when each member did not believe in it?
- Why did the Watergate participants and the administration go along with a cover-up when later testimony revealed that they all thought it was stupid at the time?

There are many examples.

The Chemical Manufacturers Association recently published a book on human error showing their concern. Here are some passages:[4]

---

4. Reprinted from *A Manager's Guide to Reducing Human Errors.* Used with permission from Chemical Manufacturers Association, Washington, DC, 1990.

During the past 5 years, about 30 major accidents at chemical and hydrocarbon processing facilities have severely injured hundreds of people, contaminated the environment, and caused more than $2 billion in property damage losses. The actual cost of these accidents was much higher because of the associated business interruption costs, cleanup costs, legal costs, fines, losses of market share, and so forth. Human error was a significant factor in almost all of these accidents. The total cost (including forced outages and off-specification products, as well as accidents) of human errors is incalculable.

Historically, managers have found human errors to be significant factors in almost every quality problem, production outage, or accident at their facilities. One study of 190 accidents in chemical facilities found the top four causes were insufficient knowledge (34%), design errors (32%), procedure errors (24%), and operator errors (16%). A study of accidents in petrochemical and refining units identified the following causes: equipment and design failures (41%), operator and maintenance errors (41%), inadequate or improper procedures (11%), inadequate or improper inspection (5%), and miscellaneous causes (2%). In systems where a high degree of hardware redundancy minimizes the consequences of single component failures, human errors may comprise over 90% of the system failure probability.

Clearly, human errors were considered the cause for an overwhelming fraction of the accidents that were evaluated in these studies. But when you stop and think about these results, they are not really surprising because there is the potential for human error in every aspect of chemical manufacturing, as indicated in Table 2-3.

In general there are two types of human errors: unintentional and intentional. Unintentional errors are actions committed or omitted with no prior thought. We typically think of these as "accidents": bumping the wrong switch, misreading a gauge, forgetting to open a valve, spilling coffee into the control console, and so forth.

Intentional deviations (errors) are actions we deliberately commit or omit because we believe, for whatever reason, that our actions are correct or that they will be better (i.e., quicker, easier, safer, etc.) than the prescribed actions. For example, workers will intentionally perform erroneous actions if they misdiagnose the true cause of an upset. Other intentional deviations are "shortcuts" that are not recognized as human errors until circumstances arise in which they exceed the system tolerances. Examples of such errors include failing to electrically ground containers of flammable liquids, attempting to restart a furnace without purging the firebox, adding a little extra catalyst to accelerate the start of a reaction, and so forth.

Human error is a natural and inevitable result of human variability in our interactions with a system. Whatever the task, human error is best understood in terms of human variability that reflects the influences of all perti-

Table 2-3. Opportunities for human error

Research/Development	The chemist failed to report that the new compound expanded when it froze
Design	The engineer failed to specify heat tracing for a heat exchanger bypass line that subsequently froze and ruptured
Construction/Installation	The contractor failed to install the specified heat tracing on a heat exchanger bypass line that subsequently froze and ruptured
Training/Procedures	The operators did not know where to turn on the heat tracing for a heat exchanger bypass line that subsequently froze and ruptured
Operation	The operators neglected their daily check of the heat tracing, which eventually failed, allowing the heat exchanger bypass line to freeze and rupture
Maintenance/Inspection	The pipefitters failed to replace the heat tracing they pinched while repairing a flange leak, allowing the heat exchanger bypass line to freeze and rupture
Plant Management	The manager delayed activation of the heat tracing system to save energy, but unpredicted cold weather froze and ruptured several pipes and vessels
Corporate Management	The corporate management cut the plant budget, forcing such severe staff reductions that the heat tracing was not all activated when cold weather arrived, resulting in a plant shutdown because of frozen lines

Source: A Manager's Guide to Reducing Human Errors, Chemical Manufacturers Association, Washington, DC, 1990.

nent factors at the time the actions are performed. There are three basic types of variability, and knowing which of these types occurs in a given case will help explain why errors occur and what can be done about them. To illustrate these types, let's consider a rifleman firing 10 shots at a target, and let's call any shot off the target an error.

1. Random variability is characterized by a dispersion pattern (Figure 2-3) centered about a desired norm—the bullseye in this example. When the variability is large, some shots will miss the target. Errors due to random variability are called *random errors*. These errors can be reduced only by reducing the overall variability of performance. Personnel selection, training, supervision, and quality control programs are ways of controlling random variability. Random errors occur when these programs are deficient, when tolerance limits are too tight, or when workers cannot control key performance factors.

2. Systematic variability is characterized by a dispersion pattern (Figure 2-4) offset from a desired norm. Although the variability may be small, the bias may cause some shots to be off target. Such errors are called *system-*

Human Error—What It Does 21

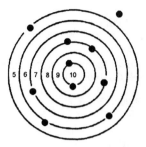

Figure 2-3. Random error. *Source: A Manager's Guide to Reducing Human Errors*, Chemical Manufacturers Association, Washington, DC, 1990.

atic errors. Bias is often the result of only one or two factors and may be easily correctable. In the example, a gun sight adjustment (or glasses) may put the rifleman right on target. Systematic errors occur, for example, when workers are given only one limit instead of a lower and an upper limit; they may deliberately attempt to be on the apparently safe (unlimited) side. Biases can also exist in tools, equipment, instructions, or the worker's personality, training, or experience. Telling workers how well they are doing with respect to real goals will help reduce systematic errors.

Anything that affects a worker's performance of a task within the process system is a *performance shaping factor* (PSF). PSFs can be divided into three classes: (1) internal PSFs that act within an individual, (2) external PSFs that act on an individual, and (3) stressors. Table 2-4 lists some internal PSFs, which are the individual skills, abilities, attitudes, and other characteristics that a worker brings to any job. Some of these, such as training, can be improved by managers; others, such as a short-term emotional upset triggered by a family crisis, are beyond any practical management control. (However, a manager's style can influence workers' mental/emotional states, as can counseling programs.)

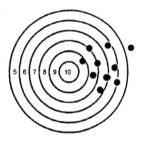

Figure 2-4. Systematic error. *Source: A Manager's Guide to Reducing Human Errors*, Chemical Manufacturers Association, Washington, DC, 1990.

Table 2-4. Internal PSFs

Training/Skill
Practice/Experience
Knowledge of Required Performance Standards
Stress (Mental or Bodily Tension)
Intelligence
Motivation/Work Attitude
Personality
Emotional State
Gender
Physical Condition/Health
Influences of Family and Other Outside Persons or Agencies
Group Identifications
Culture

Source: A Manager's Guide to Reducing Human Errors, Chemical Manufacturers Association, Washington, DC, 1990.

Table 2-5 lists external PSFs that influence the environment in which tasks are performed. These PSFs are subdivided into two groups: (1) situational characteristics and (2) task and equipment characteristics. Situational characteristics include general PSFs that may affect many different jobs in the plant. Task and equipment characteristics are pertinent to a specific job or a specific task within a job. Job and task instructions are a particularly important part of the task characteristics because they have such a large

Table 2-5. External PSFs

SITUATIONAL CHARACTERISTICS	TASK, EQUIPMENT, AND PROCEDURAL CHARACTERISTICS
Architectural Features	Procedures (Written or Not Written)
Environment	Written or Oral Communications
(Temperature, Humidity, Air Quality, Lighting, Noise, Vibration, General Cleanliness, etc.)	Cautions and Warnings
	Work Methods/Shop Practices
	Dynamic vs. Step-By-Step Activities
Work Hours/Work Breaks	Team Structure and Communication
Shift Rotation	Perceptual Requirements
Availability/Adequacy of Special Equipment, Tools, and Supplies	Physical Requirements (Speed, Strength, etc.)
	Anticipatory Requirements
Staffing Levels	Interpretation/Decision Making
Organizational Structure	Complexity (Information Load)
(Authority, Responsibility, Communication Channels, etc.)	Long- and Short-Term Memory Load
	Calculational Requirements
Actions by Supervisors, Co-Workers, Union Representatives, and Regulatory Personnel	Feedback (Knowledge of Results)
	Hardware Interface Factors
	(Design of Control Equipment, Test Equipment, Process Equipment, Job Aids, Tools, Fixtures, etc.)
Plant Policies	Control-Display Relationships
	Task Criticality
	Frequency/Repetitiveness

Source: A Manager's Guide to Reducing Human Errors, Chemical Manufacturers Association, Washington, DC, 1990.

Table 2-6. Stressor PSFs

Psychological Stressors	Physiological Stressors
Suddenness of Onset	Long Duration of Stress
High Task Speed	Fatigue
Heavy Task Load	Pain or Discomfort
High Jeopardy Risk	Hunger or Thirst
Threats (of Failure, of Loss of Job, etc.)	Temperature Extremes
Monotonous, Degrading, or Meaningless Work	Radiation
Long, Uneventful Vigilance Periods	Oxygen Deficiency
Conflicting Motives about Job Performance	Chemical Exposure
Negative Reinforcement	Vibration
Sensory Deprivation	Movement Constriction
Distractions (Noise, Glare, Movement, etc.)	Movement Repetition
Inconsistent Cueing	Lack of Physical Exercise
Lack of Rewards, Recognition, Benefits	Disruption of Circadian Rhythm

Source: *A Manager's Guide to Reducing Human Errors,* Chemical Manufacturers Association, Washington, DC, 1990.

effect on human performance. *By emphasizing the importance of preparing and maintaining clear, accurate task instructions, managers can significantly reduce the likelihood of human errors.*

The interaction between internal and external PSFs creates stress in the individual performing the task. Mismatches between our internal and external PSFs result in disruptive stress that degrades our performance. If there is too little stimulation, we will not remain sufficiently alert to do a good job. For example, an operator who passively watches as a computer operates the process is unlikely to be alert enough to take control and continue operation when the computer fails. On the other hand, too much stimulation will quickly overburden us and degrade our performance. In such situations, we tend to (1) focus on the largest or most noticeable signals and ignore some information entirely, (2) omit or delay some responses, (3) process information incorrectly and reject information that conflicts with our diagnosis or decision, or (4) mentally and/or physically withdraw. Disruptive psychological and physiological stressors are listed in Table 2-6.

REFERENCES

Bond, N. *Aviation Psychology.* Los Angeles: University of Southern California Press, 1968.

Chemical Manufacturers Association, *A Manager's Guide to Reducing Human Errors,* Washington, DC, 1990.

Harvey, J. *The Abilene Paradox.* Lexington, MA: Lexington Books, 1988.

Kletz, T., *An Engineer's View of Human Error,* the Institution of Chemical Engineers, Rugby, England, 1985.

McFarland, R. Application of human factors engineering to safety engineering problems. In J. Widener, Ed., *Selected Readings in Safety.* Macon, GA: Academy Press, 1973.

Nuberg, A. Reducing damage accidents. In J. Widener, Ed., *Selected Readings in Safety.* Macon, GA: Academy Press, 1973.

Recht, J. L. *Systems Safety Analysis: Error Rates and Costs.* Chicago: National Safety Council, 1970 Reprint.

Chapter 3

Human Error—Causes

This chapter considers the causes of human error, continuing an examination begun in the preceding chapters. You cannot define human error and talk about its characteristics and what it does to us without asking why. This chapter begins the process of asking why—why do people screw up?

In the first edition of this book the problem was finding answers to this question—there was little written although some research had been done. Today there is more research and much more written—the problem is in sorting it all out.

INFLUENCES

We start with a discussion from the first edition. Two Australians, K. B. Cooper, an accountant, and S. V. Volard, a lecturer in the Department of Management at the University of Queensland, writing in the Canadian journal *Accident Prevention*, identified the individual's contribution to the accident:[1]

> The literature on this subject can be divided into three broad categories:
>
> The influence of the work environment on the individual.
> The influence of the individual's physical characteristics.
> The influence of the individual's psychological characteristics.

---

1. Reprinted with permission from K. B. Cooper and S. V. Volard, The influence of the individual on industrial accidents, *Accident Prevention*, Toronto, Industrial Accident Prevention Association, 1978.

25

THE INFLUENCE OF THE WORK ENVIRONMENT

Under this heading attention is directed at those factors which influence the psychological climate. Psychologists have devoted a great deal of effort to examining how performance varies as external conditions change. Behind such studies is the idea that there is an optimal condition for performing each kind of task.

Stress appears to arise whenever there is a departure from optimal conditions which the individual is unable to control. Furthermore, considerations of stress have not been restricted to its effect on performance. As a worker makes more effort to maintain a given output in less than optimal conditions, this sometimes leads to accidents.

The variables considered by the safety literature and which can be classified under this heading of work environment include:

Thermal stress: for example, the combined effects of air temperature, humidity, air movement and radiant heat exchange.
Illumination: including such matters as light intensity, brightness, contrast, distribution of light, and so on.
Noise.
Hours of work.

THE INFLUENCE OF HIS PHYSICAL CHARACTERISTICS

Here, the literature looks at the various physical characteristics of workers to determine their importance in industrial safety.

The influence of the nature of the task must also be considered since on one job a variable may be important, and on another it may not. For example, good visual acuity may be important to safety in bus driving, but may have little relevance for pick and shovel work.

The range of variables which have been studied and fit the heading of physical characteristics include:

Age.
Experience.
Sex, that is, the possible differences in accident susceptibility between males and females.
Fitness, health and physical defects.
Alcoholism and drug taking.

This article is a number of years old, but is still quoted in this edition because it was then and still is one of the few safety articles to look seriously at the human side of the safety problem. The authors wrote about stress and its relationship to accidents. Today we are terribly concerned with stress, for research has shown us that people under stress do experience four to five times as many injuries as those not in stressful situations, that stress also causes psychological problems, which we are paying for, and that job stress also causes physical illness, which is covered under the workers' compensation system. Also there are many more stresses today than these authors identified, making the problem worse.

The article continues:[2]

THE INFLUENCE OF HIS PSYCHOLOGICAL CHARACTERISTICS

The variables of a psychological nature which have been studied include:

Intelligence.
Perception and motor ability.
Personality.
Emotional factors.

Several of these areas have fallen into disfavor as items for study with the loss in credibility of "accident proneness." It is now generally agreed that the small group of accident repeaters is a constantly shifting one over time, with workers dropping out of the accident group and new ones coming in. However, as Hale and Hale point out in discussing, for example, sensory-motor ability:

"The correlations were not overwhelmingly strong, but they were too large to be dismissed without due consideration."

They suggest that many of the differences in findings could be explained in terms of differences in the tests given and the jobs studied. Consequently, there is a case for examining the findings in this area, while recognizing that the demands of the job include extremely important variables.

Obviously, more studies need to be carried out in order to clarify those situations where a particular psychological characteristic is relevant to a person's safety and where it is not.

INTERRELATIONSHIPS

Many of the studies in industrial safety single out only one or two variables for investigation. This has allowed the convenient categorization of the studies into the three main areas above.

In reviewing the literature, it becomes obvious that many variables interrelate both between and within the major headings.

To illustrate these interrelationships, the variable of temperature will be

---

2. Ibid.

briefly discussed. Where temperature is the independent variable it has been shown to have the following effect:

It influences the physical discomfort of workers with the likely result of causing distraction.
In extreme situations it increases accident severity.
Hot temperature induces stiffness in the hands, thus interfering with coordination and movement. However, even though the temperature is extremely cold, the normal expectation of a greater number of accidents may not be realized if the coldness induces the wearing of safety equipment such as gloves.

Temperature can also be the dependent variable since its effects are reduced or increased by a number of factors:

The type of job will determine the extent of discomfort, since heavy manual tasks will be more adversely affected.
Both health and motivation influence the level at which a person becomes aware of discomfort and resents his environment.
Age has been shown to interrelate with temperature in some tasks, such that "older" workers have more accidents than "younger" workers when the temperature is very high.

Although a number of interrelationships exist, there are many more "empty squares." Some of these unknown interrelationships have been:

Studied, but not reviewed by the present writers.
Studied, but no relationship found and therefore not reported.
Finally, they may not have been studied at all in the past.

THE MULTIVARIATE APPROACH

As temperature is increased, older persons may [be] expected to incur relatively more accidents than younger persons.

But it may be even more useful to look at the different results achieved where the work is paced and then unpaced; in jobs which are heavy and those which are light; where there is noise and no noise, at different lighting levels and so on.

Such a study may find that the older persons' accident rate is no different where temperature is increased to 90 degrees Fahrenheit, the work is unpaced, the job is light, and the noise level is below 90 decibels.

It may also find that by changing one variable, such as unpaced to paced work, the accident liability is significantly increased.

This suggests a more realistic, if greatly more complex, method of carrying out studies into industrial safety. The hypothesis on which this method is based is that the simultaneous study of many variables will lead to the discovery of a pattern characteristic of low accident occurrence.

The number of reported studies which have used this method are, however, few.

Obviously, there is much more work to be done by practitioners and researchers alike in undertaking further study into new variables, the interrelationships between variables and multivariable analysis.

Poor industrial safety records are similar to poor performance records. They basically result from a mismatch between the demands of the job on the individual and the capacity or resources of the individual to perform his job.

The individual, therefore, cannot be overlooked in any safety program if it is to be successful.

THE HUMAN-ERROR CAUSATION MODEL

Cooper and Volard write about the environment and human characteristics (both physical and psychological) as factors that contribute to accidents and to human error, and in fact they are such factors. These authors also discussed briefly the concept of overload—of the individual being subjected to more than he or she is capable of handling. Many of these ideas are common to the field of human factors engineering.

The model of accident causation that was used in the first edition is shown in Figure 3-1.

The model was developed by the author as a result of inputs from many people. It was based on a human factors model of accident causation developed at the University of Arizona by Dr. Russell Ferrell, professor of human factors. From this base, a different model was developed for presentation to the State of the Art conference in Safety Management Concepts held by the National Safety Management Society in Washington, D.C., in December 1977. The model was then restructured into the one shown in Figure 3-1.

The model states that an injury or other type of financial loss to the company is the result of an accident or incident. The incident is the result of (1) a systems failure and (2) a human error. Systems failure is concerned with most questions traditional safety management might cover, such as:

What is management's statement of policy on safety?
Who is designated as responsible and to what degree?
Who has what authority, and the authority to do what?

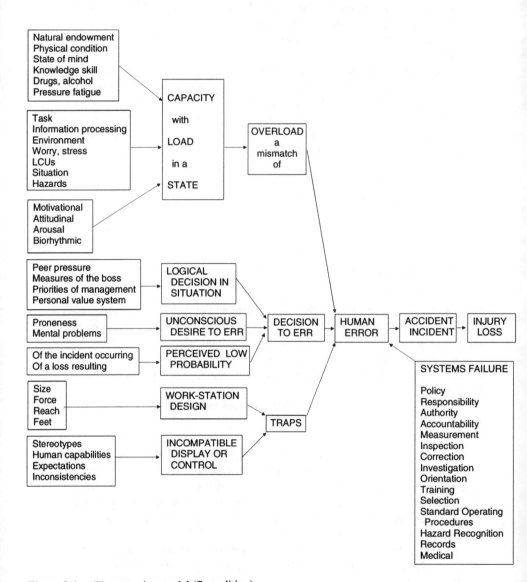

Figure 3-1. The causation model (first edition).

Who is held accountable? How?
How are those responsible measured for performance?
What systems are used for inspections to find out what went wrong?
What systems are used to correct things found wrong?
How are new people oriented?
Is sufficient training given?
How are people selected?
What are the standard operating procedures? What standards are used?
How are hazards recognized?
What records are kept, and how are they used?
What is the medical program?

These items are discussed in Part II of this book.

The second and always present aspect and cause of an incident or accident is human error. Human error results from one or a combination of three things: (1) overload, which is defined as a mismatch between a person's capacity and the load we place on him or her in a given state; (2) a decision to err; and (3) traps that are left for the worker in the workplace.

Overload

The human being cannot help but err if given a heavier work load than he or she has the capacity to handle. This overload can be physical, physiological, or psychological. To deal with overload as an accident cause, we have to look at an individual's capacity, work load, and current state. To deal with overload as an organizational cause, we have to identify the safety controls available for dealing with capacity, work load, and state.

A human being's capacity refers to physical, physiological, and psychological endowments (what the person is naturally capable of); current physical condition (and physiological and psychological condition); current state of mind; current level of knowledge and skill relevant to the task at hand; and temporarily reduced capacity owing to drugs or alcohol use, pressure, fatigue, and so on.

Load refers to the task and what it takes physically, physiologically, and psychologically to perform it. Load also refers to the amount of information processing the person must do; the working environment; the amount of worry, stress, and other psychological pressure; and the person's home life and total life situation. Load refers to a person's work situation per se and to hazards he or she faces daily at work. State refers to a person's level of motivation, attitude, arousal, and to his or her biorhythmic state.

Decision to Err

In some situations it seems logical to the worker to choose the unsafe act. Reasons for this might be as follows:

1. Because of the worker's current motivational field it makes a lot more sense to operate unsafely than safely. Peer pressure, the boss's pressure to produce, and many other reasons might make unsafe behavior seem preferable.
2. Because of the worker's mental condition, it serves him or her to have an accident. (This is called *proneness.*)
3. The worker just does not believe he or she will have an accident. (This is called *low perceived probability.*)

Traps

The third cause of human error is the traps that are left for the worker. Here we primarily are discussing human factor concepts. One trap is incompatibility. The worker errs because the situation he or she works in is incompatible with the worker's physique or with what he or she is used to. The second trap is the design of the workplace—it is conducive to human error.

The model of today (Figure 3-2) is substantially changed from that of the first edition in many ways. Instead of a simple chain of events model (A causes B which causes C, etc.), we have transformed it into more of a fault tree type of configuration to be better attuned to safety technology. And we have added a number of factors that we either were not previously aware of, or that were not present. (A lot happened in the 1980s!)

In this newer model we tried to incorporate what we have learned since the first edition, which seems to be the most important point of all—that everything we do to reduce human error, and thus losses, is totally dependent upon employee perception of the culture of the organization.

The new model incorporates this as its overriding thought: culture is the key, or, if you prefer, perception of culture (which by definition is culture) is what makes or breaks safety—what makes any element of the process work or fail.

We explain the new model this way: Management—through its vision, its honest real values, its systems of measurement and reward, and its daily decisions—creates the culture of organization. In that culture a number of processes and procedures attempt to operate. Their intent is to control losses. Some are overall (systemwide), such as whether or not people are held accountable for performance, etc.; and some are specific, such as how you select, the training a supervisor gets,

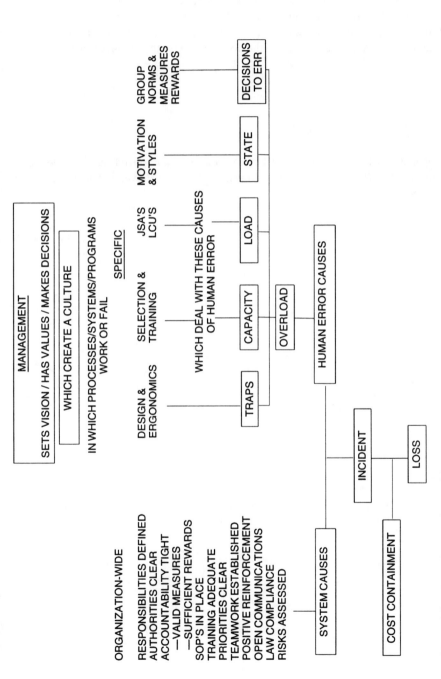

Figure 3-2. Today's causation model.

etc. These processes or procedures attempt to build a system that prevents loss, and to reduce human error due to overload, deciding to err, and traps.

When they work, there are no incidents. When they fail, one occurs. The amount of loss is determined by the seriousness of the incident (largely luck) and by the systems we have in place to control the dollar outgo (light duty programs, etc.).

REFERENCES

Cooper, K. and S. V. Volard. The influence of the individual on industrial accidents. In *Accident Prevention.* Toronto: Industrial Accident Prevention Association, 1978.

PART II | *System-Caused Human Error*

Chapter *4*

Management Causes

The causation model indicates that one of the reasons for accidents and incidents is human error and that the other reason is a systems failure.

There are a number of excellent references about systems failure if the reader is interested in pursuing this further. Most of the recent textbooks about safety management concentrate on systems failure, whether they use that terminology or not. In addition, there are a number of systems safety textbooks that also are valuable sources for the interested reader.

In our context, the concept of systems failure encompasses all reasons for accidents except reasons that have to do with the individual who has erred. It includes the failure of management to identify and correct the hazard; it includes the whole range of unsafe conditions; and it includes the lack of systems, standards, and procedures needed to deal with accident causes. Systems failure here refers to the system's failure to provide the needed safety measures.

A portion of the management oversight and risk tree (MORT) deals with this under the title Safety Program Review (Figure 4-1). While these more sophisticated approaches will be used some in large industries or government, chances are that simpler techniques are more commonly used in most industries.

A typical systems safety tool is the simple outline that follows, which was put together for use by a smaller, single-location company:

I. MANAGEMENT ORGANIZATION
   A. Does the company have a written policy on safety?
   B. Draw an organizational chart, and determine the line and staff relationships.
   C. To what extent does executive management accept its responsibility for safety?

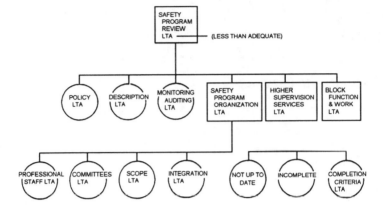

Figure 4-1. Safety program review fault tree. (From W. G. Johnson, *The Management Oversight Risk Tree—MORT*. Washington, DC: U.S. Government Printing Office, 1973.)

    1. To what extent does it participate in the effort?
    2. To what extent does it assist in administering?
  D. To what extent does executive management delegate safety responsibility? How is this accepted by:
    1. The superintendent or top production people?
    2. The foremen or supervisors?
    3. The staff safety people?
    4. The employees?
  E. How is the company organized?
    1. Is there staff safety personnel? If so, are the duties clear? Are responsibilities and authorities clear? Where is staff safety located? What can it reach? What influence does it have? To whom does it report?
    2. Are there safety committees?
      a. What is the makeup of the committees?
      b. Are the duties clearly defined?
      c. Do they seem to be effective?
    3. What type of responsibility is delegated to the employees?
  F. Does the company have written operating rules or procedures?
    1. Is safety covered in these rules?
      a. Is it built into each rule, or are there separate safety rules?
II. ACCOUNTABILITY FOR SAFETY
  A. Does management hold line personnel accountable for accident prevention?
  B. What techniques are used to fix accountability?
    1. Accountability for results:
      a. Are accidents charged against departments?

Management Causes

        b. Are claim costs charged?
        c. Are premiums prorated by losses?
        d. Does supervisory appraisal of supervisors include looking at their accident records? Are bonuses influenced by accident records?
    2. Accountability for activities:
        a. How does management ensure that supervisors conduct toolbox meetings, inspections, accident investigations, regular safety supervision, and coaching?
        b. Other?
  C. Are any special systems set up?
    1. SCRAPE
    2. Other

III. SYSTEMS TO IDENTIFY PROBLEMS—HAZARDS
  A. Are routine inspections accomplished?
    1. Who is responsible for inspection functions?
    2. By whom are inspections made?
    3. How often are they made?
    4. What types are they?
    5. To whom are the results reported?
    6. What type of follow-up action is taken?
    7. By whom?
  B. Are any special inspections made?
    1. Boilers, elevators, hoists, overhead cranes, chains and slings, ropes, hooks, electrical insulation and grounding, special machinery such as punch presses, X ray, emery wheels, ladders, scaffolding and planks, lighting, ventilation, plant trucks and vehicles, materials handling equipment, fire and other catastrophe hazards, noise and toxic controls.
  C. Are any special systems set up?
    1. Job safety analysis
    2. Critical incident technique
    3. High-potential accident analysis
    4. Fault-tree analysis
    5. Safety sampling
  D. What procedure is followed to ensure the safety of new equipment, materials, processes, or operations?
  E. Is safety considered by the purchasing department in its transactions?
  F. When corrective action is needed, how is it initiated and followed up?
  G. When faced with special or unusual jobs, how does the company ensure safe accomplishment?
    1. Is there adequate job and equipment planning?

2. Is safety a part of the overall consideration?
H. What are the normal exposures for which protective equipment is needed?
 1. What are the special or unusual exposures for which personal protective equipment is needed?
 2. What personal protective equipment is provided?
 3. How is personal protective equipment initially fitted?
 4. What type of care maintenance program is instituted for personal protective equipment?
 5. Who enforces the wearing of such equipment?

IV. SELECTION AND PLACEMENT OF EMPLOYEES
 A. Is an application blank filled out by prospective employees?
  1. Does it ask the right questions?
 B. What type of interview and screening process is the prospective employee subjected to before being hired?
 C. How are his references and past history checked?
 D. Who actually does the final hiring?
 E. Is the physical condition of the employee checked before hiring?
  1. If a physical exam is given, how complete an examination is it?
  2. How is the information used?
 F. Are any skill, knowledge, or psychological tests given?
 G. Are job physical requirements specified from job analysis?
  1. Are these requirements considered in new hires?
  2. In job transfers?

V. TRAINING AND SUPERVISION
 A. Is there safety indoctrination for new employees?
  1. Who conducts it?
  2. What does it consist of?
 B. What is the usual procedure followed in training a new employee for a job?
  1. Who does the training?
  2. How is it done?
  3. Are written job instructions based on job analysis used?
  4. Do they include safety?
 C. What training is given to an older employee who has transferred to a new job?
 D. What methods are used for training the supervisory staff?
  1. New supervisors?
  2. Continuous training for the entire supervisory force?
  3. Who does the training?
  4. Is safety a part of it?
 E. After an employee has completed the training phases of his job, what then is his status?

Management Causes 41

      1. What is the quality of the supervision?
      2. What use is made of the probation period?
VI. MOTIVATION
    A. What ongoing activities are aimed at motivation?
        1. Group meetings, literature distribution, contests, film showings, posters, bulletin boards, letters from management, incentives, house organs, accident facts on plant operations, other gimmicks and gadgets, and activities in off-the-job safety.
    B. What special emphasis campaigns have been used?
VII. ACCIDENT RECORDS AND ANALYSIS
    A. What injury records are kept? By whom?
    B. Are standard methods of frequency and severity recording used?
    C. Who sees and uses the records?
    D. What type of analysis is applied to the records?
        1. Daily analysis
        2. Weekly
        3. Monthly
        4. Annual
        5. By department
        6. Cost
        7. Other
    E. What is the accident investigation procedure?
        1. What circumstances and conditions determine which accidents will be investigated?
        2. Who does the investigating?
        3. When is it done?
        4. What type of reports are submitted?
        5. To whom do they go?
        6. What follow-up action is taken?
        7. By whom?
    F. Are any special techniques used?
        1. Estimated costs
        2. Safe-T-Scores
        3. Statistical control charts
VIII. MEDICAL PROGRAM
    A. What first aid facilities, equipment, supplies, and personnel are available to all shifts?
    B. What are the qualifications of the people responsible for the first aid program?
    C. Is there medical direction to the first aid program?
    D. What is the procedure followed in obtaining first aid assistance?
    E. What emergency first aid training and facilities are provided when normal first aid people are not available?

F. Are there any catastrophe or disaster plans?
G. What facilities are available for transportation of the injured to hospital?
H. Is a directory of qualified physicians, hospitals, ambulances, etc., available?
I. Does the company have any special preventive medicine program?
J. Does the company engage in any activities in the health education field?

Management Safety Systems

Auditing

The general term auditing refers to a number of ways that have been developed to look in detail at the safety system in operation in an organization. One such auditing system has been provided by Roman Diekemper and Donald Spartz in their article "A Quantitative and Qualitative Measure of Industrial Safety Activities." This article presents a systematic technique for the measurement of safety activity. The authors point out that there is a difference between safety activity and safety results:[1]

> While there is a relationship between the two, this relationship is nebulous at best, and requires additional research.
>
> The level and nature of activity is not always, and certainly not accurately, translated into commensurate results.
>
> Results, as expressed in both frequency and severity rates, make these measurements subject to rationalization. The same is true of cost and other measurements of results.
>
> The safety professional knows well the many and sometimes devious ways to "explain away" or even control the results expressed in various measurement terms.
>
> The logical question as to what constitutes good, valid safety activity (control) remains subject to much discussion by safety specialists. We do not attempt to answer that question, as it, too, deserves much research and continued probing.
>
> There is a need, therefore, for an objective measurement of both the quantity and quality of safety activity. The measurement of activity, after

---

1. Reprinted with permission from R. Diekemper and D. Spartz, A quantitative and qualitative measurement of industrial safety activities, *ASSE Journal*, December 1970.

all, is a way of determining how well hazards are being controlled—hopefully, through elimination. This article will deal solely with a technique for measuring the nature and level of efforts (activity) applied to control industrial hazards.

Before any measurement device can be used with reasonable accuracy to evaluate safety activities, there must be a set of rules or standards developed; that is, the measurement device must be standardized and capable of objective and continued application. We must use the same "yard stick," and use it the same way, each time an evaluation or measurement is made.

Secondly, the thing to be measured must possess certain characteristics. It must be structured so as to be measurable.

This is vital if an evaluation of safety activity is to be made at various locations of a multi-location corporation. Each plant or location must be structured with the same ground rules; i.e., a uniform "safety program."

This "program" must clearly define the structure and nature of expected activity. The responsibilities of the various management levels must be defined, and essential, basic activities (standards) which are to be implemented must be outlined; for example, accident and/or incident investigation, analysis of date for causal factors, internal self-inspection programs, etc. (Table 4-1).

Third, the measurement technique must be designed so that the line-managers can personally relate their activities to the standard.

The weighted values incorporated into the measurement device must realistically reflect the degree of importance placed by management on each of the various safety activities to be measured. This is demonstrated in the measurement technique which follows later, and is determined by in-depth analysis of past occurrences.

The criteria and the weighted values should be reviewed and changed to reflect progress. (As the application of a sound safety program becomes more sophisticated, the need to change the criteria and the weighted values increases). Again, it must be stressed that for valid comparative purposes the same set of criteria and values must be used throughout the organization.

The following measurement technique could be used for a single-plant operation or with a multi-plant operation, where the level of safety activity is comparable and a standard corporate safety "program" exists, it also lends itself to the evaluation of safety activities where the degree of sophistication is not far advanced. This would apply, unfortunately, to a wide area of American industry.

The mechanics are simple, and the process is easily understood by plant management. The format consists of three parts: (1) the activity standards, (2) a rating form, and (3) the summary sheet to compute the final score.

Table 4-1. Sample measurement technique

ACTIVITY STANDARDS

A. ORGANIZATION & ADMINISTRATION

Activity	Poor	Fair	Good	Excellent
1. Statement of policy, responsibilities assigned.	No statement of Loss Control policy. Responsibility and accountability not assigned.	A general understanding of Loss Control, responsibilities and accountability, but not written.	Loss Control Policy and responsibilities written and distributed to supervisors.	In addition to "Good" Loss Control policy is reviewed annually and is posted. Responsibility and accountability is emphasized in supervisory performance evaluations.
2. Safe operating procedures (SOP's).	No written SOP's.	Written SOP's for some, but not all hazardous operations.	Written SOP's for all hazardous operations.	All hazardous operations covered by a procedure, posted at the job location, with an annual documented review to determine adequacy.
3. Employee selection and placement.	Only pre-employment physical examination given.	In addition, an aptitude test is administered to new employees.	In addition to "Fair" new employees' past safety record is considered in their employment.	In addition to "Good" when employees are considered for promotion, their safety attitude and record is considered.
4. Emergency and disaster control plans.	No plan or procedures.	Verbal understanding on emergency procedures.	Written plan outlining the minimum requirements.	All types of emergencies covered with written procedures. Responsibilities are defined with backup personnel provisions.
5. Direct management involvement.	No measurable activity.	Follow-up on accident problems.	In addition to "Fair," management reviews all injury and property damage reports and holds supervision accountable for verifying firm corrective measures.	In addition to "Good" reviews all investigation reports. Loss Control problems are treated as other operational problems in staff meeting.
6. Plant safety rules.	No written rules.	Plant safety rules have been developed and posted.	Plan safety rules are incorporated in the plant work rules.	In addition, plant work rules are firmly enforced and updated at least annually.

Management Causes

Table 4-1. (continued)

	ACTIVITY STANDARDS			
B. INDUSTRIAL HAZARD CONTROL				
Activity	Poor	Fair	Good	Excellent
1. Housekeeping-storage of materials, etc.	Housekeeping is generally poor. Raw materials, items being processed and finished materials are poorly stored.	Housekeeping is fair. Some attempts to adequately store materials are being made.	Housekeeping and storage of materials are orderly. Heavy and bulky objects well stored out of aisles, etc.	Housekeeping and storage of materials are ideally controlled.
2. Machine guarding.	Little attempt is made to control hazardous points on machinery.	Partial, but inadequate or ineffective, attempts at control are in evidence.	There is evidence of control which meets applicable Federal and State requirements, but improvements may still be made.	Machine hazards are effectively controlled to the extent that injury is unlikely. Safety of operator is given prime consideration at time of process design.
3. General area guarding.	Little attempt is made to control such hazards as unprotected floor openings, slippery or defective floors, stairway surfaces, inadequate illuminations, etc.	Partial, but inadequate or ineffective maintenance.	There is evidence of control which meets applicable Federal and State requirements—but further improvement may still be made.	These hazards are effectively controlled to the extent that injury is unlikely.
4. Maintenance of equipment, guards, handtools, etc.	No systematic program of maintaining guards, handtools, controls and other safety features of equipment, etc.	Partial, but inadequate or ineffective maintenance.	Maintenance program for equipment and safety features is adequate. Electrical handtools are tested and inspected before issuance, and on a routine basis.	In addition to "Good" a preventive maintenance system is programmed for hazardous equipment and devices. Safety reports, files and safety department consulted when abnormal conditions are found.

Table 4-1. (continued)

	ACTIVITY STANDARDS			
Activity	Poor	Fair	Good	Excellent
5. Material handling—hand and mechanized.	Little attempt is made to minimize possibility of injury from the handling of materials.	Partial but inadequate or ineffective attempts as control are in evidence.	Loads are limited as to size and shape for handling by hand, and mechanization is provided for heavy or bulky loads.	In addition to controls for both hand and mechanized handling adequate measures prevail to prevent conflict between other workers and material being moved.
6. Personal protective equipment—adequacy and use.	Proper equipment not provided or is not adequate for specific hazards.	Partial but inadequate or ineffective provision, distribution and use of personal protective equipment.	Proper equipment is provided. Equipment identified for special hazards, distribution of equipment is controlled by supervisor. Employee is required to use protective equipment.	Equipment provided complies with standards. Close control maintained by supervision. Use of safety equipment recognized as an employment requirement. Injury record bears this out.
C. FIRE CONTROL AND INDUSTRIAL HYGIENE				
1. Chemical hazard control references.	No knowledge or use of reference data.	Data available and used by foremen when needed.	In addition to "Fair" additional standards have been requested when necessary.	Data posted and followed where needed. Additional standards have been promulgated, reviewed with employees involved and posted.
2. Flammable and explosive materials control.	Storage facilities do not meet fire regulations. Containers do not carry name of contents. Approved dispensing equipment not used. Excessive quantities permitted in manufacturing areas.	Some storage facilities meet minimum fire regulations. Most containers carry name of contents. Some appoved dispensing equipment in use.	Storage facilities meet minimum fire regulations. Most containers carry name of contents. Approved equipment generally is used. Supply at work area is limited to one day requirement. Containers are kept in approved storage cabinets.	In addition to "Good" storage facilities exceed the minimum fire regulations and containers are always labeled. A storage policy is in evidence relative to the control of the handling, storage and use of flammable materials.
3. Ventilation—fumes, smoke and dust control.	Ventilation rates are below industrial hygiene standards in areas where there is an industrial hygiene exposure.	Ventilation rates in exposure areas meet minimum standards.	In addition to "Fair" ventilation rates are periodically measured, recorded and maintained at approved levels.	In addition to "Good" equipment is properly selected and maintained close to maximum efficiency.

Table 4-1 (continued)

		ACTIVITY STANDARDS		
Activity	Poor	Fair	Good	Excellent
4. Skin contamination control.	Little attempt at control or elimination of skin irritation exposures.	Partial, but incomplete program for protecting workers. First-aid reports on skin problems are followed up on an individual basis for determination of cause.	The majority of workmen instructed concerning skin-irritating materials. Workmen provided with approved personal protective or devices. Use of this equipment is enforced.	All workmen informed about skin-irritating materials. Workmen in all cases provided with approved personal protective equipment or devices. Use of proper equipment enforced and facilities available for maintenance. Workers are encouraged to wash skin frequently. Injury record indicates good control.
5. Fire control measures.	Do not meet minimum insurance or municipal requirements.	Meets minimum requirements.	In addition to "Fair" additional fire hoses and/or extinguishers are provided. Welding permits issued. Extinguishers on all welding carts.	In addition to "Good" a fire crew is organized and trained in emergency procedures and in the use of fire fighting equipment.
6. Waste-trash collection and disposal, air/water pollution.	Control measures are inadequate.	Some controls exist for disposal of harmful wastes or trash. Controls exist but are ineffective in methods or procedures of collection and disposal. Further study is necessary.	Most waste disposal problems have been identified and control programs instituted. There is room for further improvement.	Waste disposal hazards are effectively controlled. Air/water pollution potential is minimal.

D. SUPERVISORY PARTICIPATION, MOTIVATION AND TRAINING

1. Line supervisor safety training.	All supervisors have not received basic safety training.	All shop supervisors have received some safety training.	All supervisors participate in division safety training session, a minimum of twice a year.	In addition, specialized sessions conducted on specific problems.
2. Indoctrination of new employees.	No program covering the health and safety job requirements.	Verbal only.	A written handout to assist in indoctrination.	A formal indoctrination program to orientate new employees is in effect.
3. Job hazard analysis.	No written program.	Job hazard analysis program being implemented on some jobs.	JHA conducted on majority of operations.	In addition, job hazard analyses performed on a regular basis and safety procedures written and posted for all operations.

Table 4-1 (continued)

		ACTIVITY STANDARDS		
Activity	Poor	Fair	Good	Excellent
4. Training for specialized operations (Fork trucks, grinding, press brakes, punch presses, solvent handling, etc.)	Inadequate training given for specialized operations.	An occasional training program given for specialized operations.	Safety training is given for all specialized operations on a regular basis and retraining given periodically to review correct procedures.	In addition to "Good" an evaluation is performed annually to determine training needs.
5. Internal self-inspection.	No written program to identify and evaluate hazardous practices and/or conditions.	Plant relies on outside sources, i.e., insurance safety engineer and assumes each supervisor inspects his area.	A written program outlining inspection guidelines, responsibilities, frequency and follow up is in effect.	Inspection program is measured by results, i.e., reduction in accidents and costs. Inspection results are followed up by top management.
6. Safety promotion and publicity.	Bulletin boards and posters are considered the primary means for safety promotion.	Additional safety displays, demonstrations, films, are used infrequently.	Safety displays and demonstrations are used on a regular basis.	Special display cabinets, windows, etc. are provided. Displays are used regularly and are keyed to special themes.
7. Employee/supervisor safety contact and communication.	Little or no attempt made by supervisor to discuss safety with employees.	Infrequent safety discussion between supervisor and employees.	Supervisors regularly cover safety when reviewing work practices with individual employees.	In addition to items covered under "Good" supervisors make good use of the shop safety plan and regularly review job safety requirements with each worker. They contact at least one employee daily to discuss safe job performance.
E. ACCIDENT INVESTIGATION, STATISTICS AND REPORTING PROCEDURES				
1. Accident investigation by line personnel.	No accident investigation made by line supervision.	Line supervision makes investigations of only medical injuries.	Line supervision trained and makes complete and effective investigations of all accidents; the cause is determined; corrective measures initiated immediately with a completion date firmly established.	In addition to items covered under "Good" investigation is made of every accident within 24 hours of occurrence. Reports are reviewed by the department manager and plant manager.
2. Accident cause and injury location analysis and statistics.	No analysis of disabling and medical cases to identify prevalent causes of accidents and location where they occur.	Effective analysis by both cause and location maintained on medical and firstaid cases.	In addition to effective accident analysis, results are used to pinpoint accident causes so accident prevention objectives can be established.	Accident causes and injuries are graphically illustrated to develop the trends and evaluate performance. Management is kept informed on status.

Table 4-1. (continued)

ACTIVITY STANDARDS

Activity	Poor	Fair	Good	Excellent
3. Investigation of property damage.	No program.	Verbal requirement or general practice to inquire about property damage accidents.	Written requirement that all property damage accidents of $50 and more will be investigated.	In addition, management requires a vigorous investigation effort on all property damage accidents.
4. Proper reporting of accidents and contact with carrier.	Accident reporting procedures are inadequate.	Accidents are correctly reported on a timely basis.	In addition to "Fair" accident records are maintained for analysis purposes.	In addition to "Good" there is a close liaison with the insurance carrier.

RATING FORM

	Poor	Fair	Good	Excellent	Comments
A. ORGANIZATION & ADMINISTRATION					
1. Statement of policy, responsibilities assigned.	0	5	15	20	
2. Safe operating procedures (SOP's).	0	2	15	17	
3. Employee selection and placement.		2	10	12	
4. Emergency and disaster control planning.	0	5	15	18	
5. Direct management involvement.	0	10	20	25	
6. Plant safety rules.	0	2	5	8	
Total value of circled numbers	___	+ ___	+ ___	+ ___ X .20 Rating ___	
B. INDUSTRIAL HAZARD CONTROL					
1. Housekeeping—storage of materials, etc.	0	4	8	10	
2. Machine guarding.	0	5	16	20	
3. General area guarding.	0	5	16	20	
4. Maintenance of equipment guards, hand tools, etc.	0	5	16	20	
5. Material handling—hand and mechanized.	0	3	8	10	
6. Personal protective equipment—adequacy and use.	0	4	16	20	
Total value of circled numbers	___	+ ___	+ ___	+ ___ X .20 Rating ___	
C. FIRE CONTROL & INDUSTRIAL HYGIENE					
1. Chemical hazard control references.	0	6	17	20	
2. Flammable and explosive materials control.	0	6	17	20	
3. Ventilation—fumes, smoke and dust control.	0	2	8	10	
4. Skin contamination control.	0	3	10	15	
5. Fire control measures.	0	2	8	10	
6. Waste—trash collection and disposal, air/water pollution.	0	7	20	25	
Total value of circled numbers	___	+ ___	+ ___	+ ___ X .20 Rating ___	

Table 4-1. (continued)

RATING FORM	Poor	Fair	Good	Excellent	Comments
D. SUPERVISORY PARTICIPATION, MOTIVATION & TRAINING					
1. Line supervisor safety training.	0	10	22	25	
2. Indoctrination of new employees.	0	1	5	10	
3. Job hazard analysis.	0	2	8	10	
4. Training for specialized operations.	0	2	7	10	
5. Internal self-inspection.	0	5	14	15	
6. Safety promotion and publicity.	0	1	4	5	
7. Employee/supervisor contact and communication.	0	5	25	25	
Total value of circled numbers	___	+ ___	+ ___	+ ___ X .20 Rating ___	
E. ACCIDENT INVESTIGATION, STATISTICS & REPORTING PROCEDURES					
1. Accident investigation by line supervisor.	0	10	32	40	
2. Accident cause and injury location analysis and statistics.	0	3	8	10	
3. Investigation of property damage.	0	10	32	40	
4. Proper reporting of accidents and contact with carrier.	0	3	8	10	
Total value of circled numbers	___	+ ___	+ ___	+ ___ X .20 Rating ___	

SUMMARY

The numerical values below are the weighted ratings calculated on rating sheets. The total becomes the overall score for the location.

A. Organization & Administration _____

B. Industrial Hazard Control _____

C. Fire Control & Industrial Hygiene _____

D. Supervisory Participation, Motivation & Training _____

E. Accident Investigation Statistics & Reporting Procedures _____

TOTAL RATING _____

Profiling

Profiling consists of the development of a standard of corporate safety performance in a number of categories considered to be important. Companies are then compared to that standard, and a profile is made to show how the company compares to the standard in a number of categories.

Profiling is used in Canada and is described in the works of Jack Fletcher, who, in his book *The Industrial Environment,* provides a means of rating or grading an organization in the areas of injury prevention, damage control, and total loss control. For injury prevention he provides a rating outline for such items as loss control policy, guarding, inspecting, design and purchasing, audiovisual aids, committees and rules, training, investigation, records and analysis, costs, medical examinations, and personal protective equipment. Fletcher provides a way to profile the following items under total loss control: fire prevention and control, security, health and hygiene, pollution, and product integrity. He provides a system of ratings from 0 to 5, or from unsatisfactory to excellent, and provides the criteria for each rate.

James Tye, the director general of the British Safety Council, offers a similar approach in his book *Management Introduction to Total Loss Control,* in which he explains his profiling system, by which the profiler rates an organization in 30 different areas and rates safety performance on a percentage scale. His master evaluation and development grid is shown in Figure 4-2. Each of the 30 areas is broken down into key elements. For example, the key area of management involvement has the evaluation and development grid shown in Figure 4-3.

Profiling is used widely in South Africa. The National Occupational Safety Association (NOSA) (the counterpart of the National Safety Council in the United States) uses a profiling approach to grade every industry. The general manager of NOSA describes the process as follows:[2]

> The industrialist in South Africa has had much of the preparatory work done for him when it comes to Management by Objectives (MBO) in the field of Accident Prevention. Using the best expertise available. NOSA's technical staff, many years ago, arrived at the major objectives required to institute a successful safety programme. In broad terms this covered: Industrial housekeeping; electrical, mechanical and personal safeguarding; fire prevention and control; accident recording and investigation; and safety organization. Each major objective was then broken down into subsidiary objectives giving a total of 50 items, which management should strive to obtain.

---

2. Reprinted from B. Matthysen, *The NOSA Safety Effort Rating System,* Arcadia, South Africa, National Occupational Safety Association, 1975.

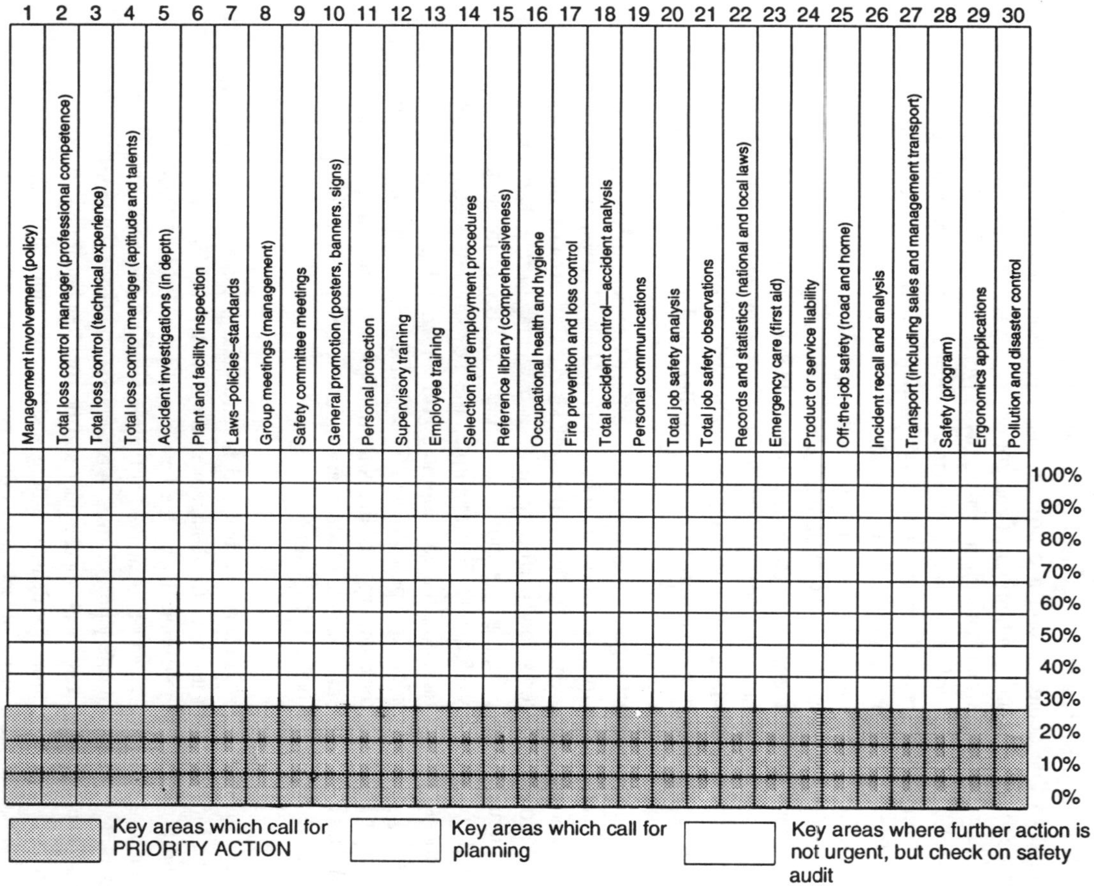

Figure 4-2. Master evaluation and development grid. (From J. Tye, Management introduction to *Total Loss Control*. London: British Safety Council, 1970.)

Management Causes 53

Key Area: Management Involvement										
Is Written Safety Policy Issued	Is Safety Policy Circulated to Top, Middle and Supervisory Management, To Shop Floor, Apprentices and New Entrants?	Is Policy Re-issued Quarterly In Some Written Form?	Are Executive Safety Inspections on Quarterly Basis?	Is Safety Policy Enforcement Effective?	Attitude of Key Management Personnel To Safety Policy	Is Management Action Safety Programme Operating Effectively?	Is An Effective Safety Performance Report made at Management Meetings?	Are Effective Safety Management Audits Executed Annually?	Does Foreman/Supervisor's Job Appraisal Include Safety Performance?	
										100%
										90%
										80%
										70%
										60%
										50%
										40%
										30%
										20%
										10%
										0%
										Total _____
Divide the total percentages by number of questions (10) to obtain average result to carry over to master evaluation and development chart. _____										

Figure 4-3. Evaluation and development of management involvement. (From J. Tye, Management introduction to *Total Loss Control*. London: British Safety Council, 1970.)

Some of the objectives were considered more important than others. So in order to differentiate them it was decided to quantify the items. The more important ones were given heavier mark allocations. For example, machine guarding carries 150 marks whereas clean premises has 40 marks allocated. The total allocation for the 50 items is 2,000 marks.

Under the guidance of the NOSA technical staff, the optimum requirements for each factory can be determined, using the criterion: HOW MUCH MORE could management reasonably be expected to do within the specific plant; taking COGNISANCE of the materials, the methods, the men and the money available to the plant.

When the safety state of a plant is to be audited by a NOSA technical staff member, the plant starts with a clean sheet and its full quota of 2,000 marks. When deviations from the set objective are observed they are discussed with management and a note made thereof. Once all 50 items have been investigated in depth the final mark allocation is made by the NOSA man.

From this unique system has evolved NOSA's Star Grading Scheme, whereby recognition for attaining the set objectives is given. To obtain a Five Star Grading the firm must have a mark allocation of over 90 per cent. But attaining the objectives must also result in a reduction in the injury frequency rate. Therefore another criterion is added to the requirements of a Five Star Grading—the injury frequency rate must not be greater than five disabling injuries per one million man-hours exposure. At the other end of the scale, in order to attain a One Star Grading, the mark allocation must be between 40 and 50 per cent and the injury frequency rate must not exceed 25.

The NOSA report form no. 4.13.1 (Figure 4-4) is sent to management after each safety audit where X marks the spot for major problem areas.

Using the NOSA MBO system management can achieve fantastic results in advertising their objectives of conserving our greatest asset—skilled man-power.

Simple checklists, sophisticated fault trees, audits, and profiles are all ways of systematically examining the system before and after accidents to find systems failures.

The above provides some examples of published audits or profiles, to give the reader an idea of different opinions as to what should be in a management safety system. We only selected a few of those that are available, ranging from a simple checklist (Figure 4-1) to a more detailed quantified audit (Table 4-1) to the profiling systems from other countries.

All include different things reflecting their authors' opinions of what the essential components of a safety system are; and these are only opinions. What are the essential elements of a safety system? We do not know. We have many pieces of research on what the essential elements are, and the research continually tells us that there are *no* essential elements of a safety system—that what works in one organization may not in another, that what works in one location of your company may not in another.

The research further suggest that it is not the elements that you choose that get results in safety, but, rather, it is the culture in which those elements live that determines whether or not they will work.

In this part of the book, we are discussing "System-Caused Human Error"

Management Causes 55

CONFIDENTIAL
SURVEY/RATING OF MESSRS.

NATIONAL OCCUPATIONAL SAFETY ASSOCIATION
SAFETY EFFORT SURVEY/RATING

Form 4.13.1

NOTE: Items marked "X" require management's attention and should be read in conjunction with the accompanying report. Please refer to the booklet "Management by Objectives" for advice on effective management practice in accident prevention.

		Max.	Actual	Action				MAX.	Actual	Action
1.00	PREMISES AND HOUSEKEEPING				4.00	ACCIDENT RECORDING AND INVESTIGATION				
	1.10 PREMISES					4.10 RECORDS				
	1.11 Buildings and floors—clean and in good state of repair	40				4.11 Adequate accident recording (Register and dressing book)		30		
	1.12 Good lighting (natural and artificial)	20				4.12 Internal accident report form signed by supervisory staff		30		
	1.13 Ventilation	30				4.13 Adequate accident statistics kept in accessible place and NOSA informed		30		
	1.20 HOUSEKEEPING					4.20 INVESTIGATION of accidents and remedial measures taken to prevent recurrence		30		
	1.21 Aisles and storage demarcated	30								
	1.22 Good stacking and storage practices.	50				SECTION RATING		150		%
	1.23 Factory and yard—clean of superfluous material	60			5.00	SAFETY ORGANIZATION				
	1.24 Scrap and refuse bins—removal and disposal	30				5.10 SAFETY PERSONNEL				
	1.25 Color coding—machines, pipelines—other	40				5.11 One person made responsible for safety coordination by management, in writing, i.e. safety officer, permanent/part-time		30		
	SECTION RATING	300		%		5.12 Appointment and acceptance of appointment in terms of Factory Regulations C.7.2(a) and (b) or Mines and Works Regulation 2.9.2		30		
2.00	ELECTRICAL, MECHANICAL AND PERSONAL SAFEGUARDING									
	2.10 MECHANICAL EQUIPMENT									
	2.11 Machine guarding	150				5.13 European Safety Committee		80		
	2.12 Lock-out system and usage	40				5.14 Non-European Safety Committee or any other similar system		40		
	2.13 Labelling of shut-off valves, switches, isolators	30				5.15 First aider and equipment		20		
	2.14 Ladders, stairs, walkways, platforms	40				5.16 First aid training		30		
	2.15 Lifting gear and records	40				5.20 SAFETY PROPAGANDA				
	2.16 Compressed gases; pressure vessels and records	30				5.21 Poster programmes, bulletins, newsletters, use of safety films and internal safety competitions, etc.		130		
	2.20 ELECTRICAL EQUIPMENT									
	2.21 Portable electrical equipment—monthly checks and records	40				5.22 Notice board indicating injury experience		20		
	2.22 Earth leakage relays—permanent and portable	30				5.23 Suggestion scheme		20		
	2.23 General electrical installation	50				5.30 INDUCTION TRAINING AND JOB INSTRUCTION, continuous training, e.g. poster appreciation, lectures, rule book. NOSA safety training courses		50		
	2.30 HAND-TOOLS—All types: condition, storage and use., e.g. hammers, chisels	50								
	2.40 PROTECTIVE EQUIPMENT (issued: use)					5.40 PLANT INSPECTION system of reporting to management on safety conditions		50		
	2.41 Head protectors	20								
	2.42 Eye protectors	20				5.50 WRITTEN SAFETY OPERATING PRACTICES and procedure issued, displayed and explained to the illiterate		50		
	2.43 Foot protectors	20								
	2.44 Protective clothing, including hand protectors	20				5.60 ANY ITEM NOT DETAILED				
	2.45 Respiratory equipment	20				5.61 COMPANY POLICY (Also Total Loss Control Programme)		100		
	2.46 Maintenance	20				5.62 Bonus points awarded or penalty points deducted				
	2.50 NOTICES—Electrical, mechanical, protective equipment, etc.	30				SECTION RATING		650		%
	SECTION RATING	650		%		OVERALL RATING				%
3.00	FIRE PROTECTION AND PREVENTION									
	3.01 Correct types of extinguishers	40								
	3.02 Areas demarcated and clear, extinguishers	20								
	3.03 Locations marked	20								
	3.04 Maintenance of equipment	30								
	3.05 Storage flammable material	30								
	3.06 Signs to exits: alarm system	30								
	3.07 Fire fighting drill and instructions on fire extinguishors	80								
	SECTION RATING	250		%						

EFFORT RATING VALUES
NO. OF STARS   RATING %
5* Excellent   91–100
4* Very good   75–90
3* Good   61–74
2* Average   51–60
1* Fair   40–50

Figure 4-4. Survey form used by NOSA. (From National Occupational Safety Association, Arcadia, South Africa.)

and in this chapter the management causes of unsafe behavior, yet we cannot prescribe what that "system" should consist of for any individual's organization. We can, however, tell when a system failed, or better yet when it is likely to fail. When people do not know what is unsafe, a system is primed for failure; if people do not care, it is primed; when hazards exist unrecognized, it is primed, and so on.

Your means, at your location, of ensuring that these things do not exist can be accomplished in a variety of ways, allowing you great flexibility and the opportunity for innovation and creativity. The checklists and audits in this chapter have worked for some people. You must decide what is right in your organization.

Assessment Criteria

While we cannot tell any organization exactly what will work for it, we can offer some solid criteria that you can judge your management system against, based upon the existing research. If your safety system:

1. Forces regular, daily supervisory action that is proactive and people-oriented,
2. Actively, daily involves your middle managers,
3. Gets top management visibility,
4. Gets active meaningful worker participation,
5. Allows flexibility, and
6. Is perceived as positive by the workers,

then your system, whatever it looks like or consists of, will get results. These six criteria are the key to safety excellence.

Keep in mind that these criteria will not assure legal compliance. OSHA's guidelines (and some states now require by law a written safety program) are filled with elements and components believed to be "right." Most of these are based on the Heinrich thinking (discussed in the Introduction) from the early days of safety.

To achieve the six criteria you will probably need some mechanisms that fit your culture to:

- Define required tasks for each level.
- Measure task completion for each level.

Management Causes

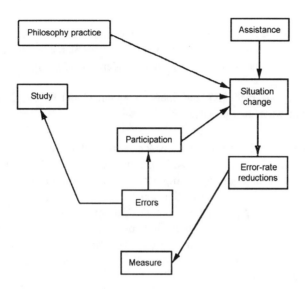

Figure 4-5. Improving human performance. (From W. Johnson, *The Management Oversight Risk Tree—MORT.* Washington, DC: U.S. Government Printing Office, 1973.)

- Reward for task completion.
- Identify system weaknesses.
- Observe and deal with human error, etc.

The choice of mechanisms is up to you; they should be in tune with the style, climate, and culture of your organization

W. G. Johnson, writing in his text *The Management Oversight Risk Tree—MORT,* provides us with an excellent starting point for our discussion on human-error reduction by providing us with six error-reduction concepts and a model for improving human performance (Figure 4-5). In Johnson's concepts,[3] we can see the same thinking as expressed earlier by Chapanis:

1. Errors are an inevitable (rate-measurable) concommitant [sic] of doing work or anything.
2. Situations may be error-provocative—changing the situation will likely do more than elocution or discipline.
3. Many error definitions are "forensic" (which is dabatable [sic], imprecise, and ineffective) rather than precise.

---

3. From W. Johnson, *The Management Oversight and Risk Tree—MORT,* Washington, DC, U.S. Government Printing Office, 1973.

4. Errors at one level mirror service deficiencies at a higher level.
5. People mirror their bosses—if management problems are solved intuitively, or if chance is relied on for non-accident records, long-term success is unlikely.
6. Conventional methods of documenting organizational procedures seem to be somewhat error-provocative.

In concept 2, Johnson brings up the notion that certain situations are "error-provocative," a notion we will examine in some detail in later chapters. This concept is perhaps the most important single contribution made by error-reduction theorists. It leads us to the conclusion expressed by Johnson in the second part of concept 2, that is, that we can make a lot more progress by changing the situation than by preaching or disciplining. Johnson's concepts 4 and 5 are also of key importance to the material in this book: human errors at lower levels of the organization are symptoms of things that are wrong in the organization at higher levels.

Johnson's model of error reduction (Figure 4-5) suggests that we can reduce human error by changing the situation. This change is accomplished by assistance from the outside (staff safety, line management, etc.), working within a corporate philosophy, through study of the situation, and through participation of the individual worker.

Besides his excellent definition. Chapanis provides some additional interesting observations:[4]

1. Many situations are error-provocative.
2. Given a population of human beings with known characteristics, it is possible to design tools, appliances, and equipment that best match their capacities, limitations, and weaknesses.
3. The improvement in system performance that can be realized from the redesign of equipment is usually greater than the gains that can be realized from the selection and training of personnel.
4. For purposes of man-machine systems design there is no essential difference between an error and an accident. The important thing is that both an error and an accident identify a troublesome situation.
5. The advantages of analyzing error-provocative situations are:
   a. It is easier to collect data on errors and near-misses than on accidents.
   b. Errors occur much more frequently than do accidents. This means, in short, that more data are available.

---

4. Reprinted with permission from A. Chapanis, in W. Johnson, *New Approaches to Safety in Industry*, London, InComTec, 1972.

Management Causes

c. Even more important than the first two points is that error-provocative situations provide one with clues about what one can do to prevent errors, or accidents, before they occur.
d. The study of errors and near-misses usually reveals all those situations that result in accidents plus many situations that could potentially result in accidents but that have not yet done so. In short, by studying error-provocative situations we can uncover dangerous or unsafe designs even before an accident has had a chance to occur. This, in fact, is one of the keys to designing safety into a system before it is built.
e. If we accept that the essential difference between an error and an accident is largely a matter of chance, it follows that any measure based on accidents alone, such as number of disabling injuries, injury frequency rates, injury severity rates, number of first-aid cases, and so on, is contaminated by a large proportion of pure error variability. In statistical terms the reliability of any measure is inversely related to the amount of random, or pure error, variance that contributes to it. It is likely that the reason so many studies of accident causation turn up with such marginally low relationships is the unstable, or unreliable, nature of the accident measure itself.

6. Design characteristics that increase the probability of error include a job, situation, or system which:
   a. violates operator expectations,
   b. requires performance beyond what an operator can deliver,
   c. induces fatigue,
   d. provides inadequate facilities or information for the operator,
   e. is unnecessarily difficult or unpleasant, or
   f. is unnecessarily dangerous.

LOGIC DIAGRAMS ON ERROR CAUSES

A number of people have provided us with logic diagrams to explain the system causes of human error. Johnson, in MORT, offers the fault tree shown in Figure 4-6.

A similar logic diagram on the causes for human error was constructed by Jorge Hernandez Osuna of Monterrey, Mexico (Figure 4-7). This logic diagram identifies three basic reasons behind unsafe acts: (1) the employee did not know what to do, (2) the employee was unable to do it safely, and (3) the employee refused to do it safely.

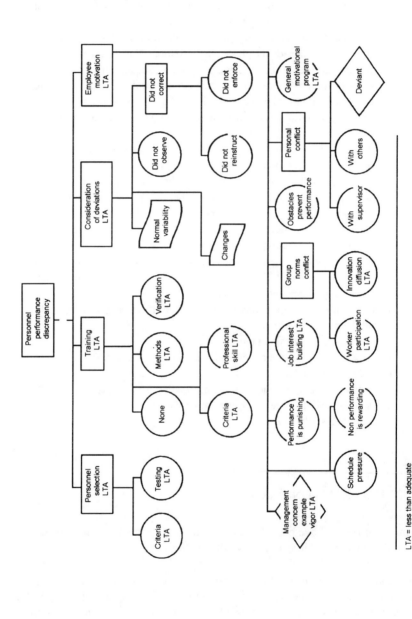

Figure 4-6. Human-error fault tree. (From W. Johnson, *The Management Oversight Risk Tree—MORT*. Washington, DC: U.S. Government Printing Office, 1973.)

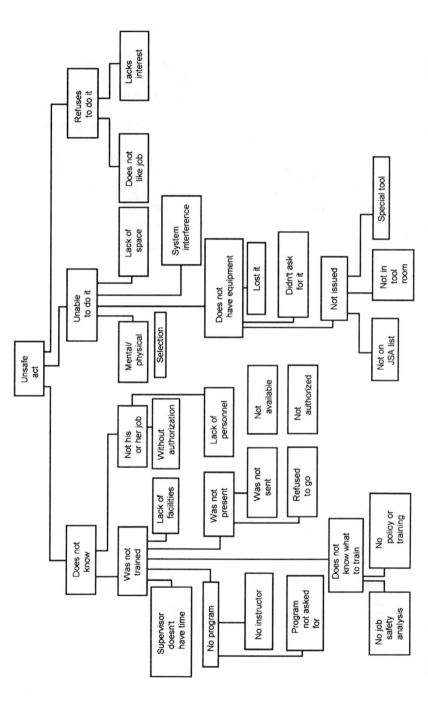

Figure 4-7. Logic diagram of the causes of human error. (From J. Osuna, Why the unsafe act? Paper read before the Industrial Accident Prevention Association of Ontario, April 1976, in Toronto.)

REFERENCES

Chapanis, A. Quoted in W. Johnson, *New Approaches to Safety in Industry.* London: InComTec, 1972.

Diekemper, R. and D. Spartz. A quantitative and qualitative measurement of industrial safety activities. *ASSE Journal,* December 1970.

Fletcher, J. *The Industrial Environment.* Willowdale, Ontario: National Profile, Ltd., 1972.

Johnson, W. *The Management Oversight and Risk Tree—MORT.* Washington, DC: U.S. Government Printing Office, 1973.

Matthysen, B. *The NOSA Safety Effort Rating System.* Arcadia, South Africa: National Occupational Safety Association, 1975.

Osuna, J. Why the unsafe act? Paper read before the Industrial Accident Prevention Association of Ontario, April 1976, in Toronto. Photocopied.

Petersen, D. A loss control analysis guide. In *Techniques of Safety Management,* 2nd ed. New York: McGraw-Hill, 1978.

Tye, J. *Management Introduction to Total Loss Control.* London: British Safety Council. 1970.

Chapter 5

Culture Causes

In Chapter 4 we considered system causes of incidents from two standpoints: (1) when the management system does not adequately deal with situations and conditions that might hurt people, and causes loss; and (2) when the management system creates an environment that makes it more likely for a person to err. Both will lead to system-caused human error and/or to incidents, accidents, and financial loss.

Outside of the field of safety, a number of people have examined the causes of human error. Typical are the psychological factors identified by Minor and Brewer as contributing to ineffective performance in business organizations:[1]

1. Intelligence and job knowledge
   a. Insufficient verbal ability
   b. Insufficient special ability other than verbal
   c. Insufficient job knowledge
   d. Defect of judgement or memory
2. Emotions and emotional illness
   a. Continuing disruptive emotion—anxiety, depression, anger, excitement, shame, guilt, jealousy
   b. Psychosis
   c. Neurosis
   d. Alcoholism or drug problems
3. Individual motivation to work
   a. Strong motives frustrated at work

---

1. Reprinted from J. Minor and J. Brewer, The management of ineffective performance, in *Handbook of Industrial and Organizational Psychology*, edited by M. Dunnette, Chicago, Rand-McNally, 1976.

b. Unintegrated means to satisfy motives
   c. Inappropriate physical characteristics
   d. Generalized low work motivation
4. Physical characteristics and disorders
   a. Physical illness or handicap, including brain damage
   b. Physical disorders of emotional origin
   c. Inappropriate physical characteristics
   d. Insufficient muscular or sensory ability
5. Family ties
   a. Family crises
   b. Separation from an emotionally significant family
   c. Social isolation
   d. Predominance of family considerations over work demands
6. The groups at work
   a. Negative consequences associated with group cohesion
   b. Ineffective management
   c. Inappropriate managerial standards or criteria
7. The company
   a. Insufficient organizational action
   b. Placement error
   c. Organizational over-permissiveness
   d. Excessive span of control
   e. Inappropriate organizational standards or criteria
8. Society and its values
   a. Application of legal sanctions
   b. Enforcement of cultural values by means not connected with the administration of the law
   c. Conflict between job demands and cultural values as individually held (equity, freedom, morality, etc.)
9. Situational forces
   a. Negative consequences of economic forces
   b. Negative consequences of geographic location
   c. Excessive danger
   d. Problems in the work itself

A number of these factors would fall under the category of organizational "cultural causes." Those listed under numbers 7, 8, and 9 are to a large degree "cultural causes" within the organization.

As indicated in Chapter 4, the culture of the organization is what determines whether or not the elements of the safety system will be effective. In effect, the culture will determine the organization's safety results. The concept of culture has been perhaps the single most important topic discussed in management theory in the last ten or fifteen years.

Culture Definition

Probably the best definition I have heard of culture came from a worker I interviewed: "Culture is what everybody knows and therefore does not have to be stated or written down."

The concept became a very popular management subject in the early 1980s, probably because of the popularity of the book *In Search of Excellence* by Peters and Waterman. That book described what it was that accounted for the economic success of a number of companies—what made them good. Other works followed, delving into the concept.

Basically the concept of culture had been considered long before that. In 1967 Dr. Rensis Likert wrote *The Human Organization,* in which he described his research on trying to understand the difference in "styles" of different companies and how these "styles" affected the bottom line. Dr. Likert coined the term "organizational climate." We now call that culture.

Likert not only researched climate; he also evaluated it in ten areas.[2]

1. Confidence and trust.
2. Interest in the subordinate's future.
3. Understanding of and the desire to help overcome problems.
4. Training and helping the subordinate to perform better.
5. Teaching subordinates how to solve problems rather than giving the answer.
6. Giving support by making available the required physical resources.
7. Communicating information that the subordinate must know to do the job, as well as information needed to identify more with the operation.
8. Seeking out and attempting to use ideas and opinions.
9. Approachability.
10. Crediting and recognizing accomplishments.

Some of Likert's ideas are discussed in Chapter 11 when we discuss where motivation fits into human error. Likert invented a way to measure climate with a forced choice questionnaire, which he administered to employees of an organization to determine their perception of how good the company was in the ten areas. He later took the perception survey results and ran correlational studies such measures as profitability, return on investment, growth, and other bottom-line figures, invariably coming up with extremely high positive correlations. Apparently climate determines results.

---

2. From R. Likert, *The Human Organization,* New York, McGraw-Hill, 1967.

As culture (climate) became a popular management subject in the 1980s, executives began to look at their organizations and consider ways to "improve their culture." Many organizations displayed new posters on the walls describing "their culture." We know today that if the management of a company must write it down and make a poster of it, they are not describing their culture—they are describing what they would like it to be. No employees need the culture to be described to them—everybody knows "what it is like around here."

CULTURE AND SAFETY

The field of safety pretty much ignored the concept of culture throughout the 1980s. As management attempted to improve culture through changing styles of leadership, employee participation, and so forth, safety professionals tended to change their approaches very little, keeping the same tools, using the same elements in their safety "programs" that they had always used. Safety programs typically consisted of the usual things: meetings, inspections, accident investigations, using job safety analyses (JSAs), and so on. These tools were perceived as the essential elements of a safety program. OSHA published a guideline in the 1980s suggesting that all companies should follow all of these practices. A number of states enacted laws requiring companies to do these things. These traditional elements were regarded as a "safety program."

While OSHA and the states were going down the "essential element" track to safety (as was much of the safety professions—see Chapter 4 on auditing), a number of researchers began to suggest totally different answers to the safety problem. Most of their research results were consistent in saying that "there are no essential elements"—what works in one organization will not in another. Each organization must determine for itself what will work for it—there are no magic pills. The answer seems to be clear: it is the culture of the organization that determines what will work in that organization.

Certain cultures do, in fact, have safety as a key component, whereas other cultures make it very clear that safety is unimportant. In the latter almost nothing will work; meetings will be boring, JSAs perceived only as paperwork, and so on.

The culture of the organization sets the tone for everything in safety. In a positive safety culture, it says that everything you do about safety is important. In a participative culture, the organization is saying to the worker, "We want and need your help." Some cultures urge creativity and innovation; some destroy it. Some cultures tap the employees for ideas and help; some force the employees never to use their brains at work.

WHAT SETS THE CULTURE

Here we mention just a few considerations that determine the culture:

- How decisions are made: Does the organization spend its available money on people? On safety? Or are these ignored for other things?
- How people are measured: Is safety measured as tightly as production? What is measured tightly is what is important to management.
- How people are rewarded: Is there a larger reward for productivity than for safety? This states management's real priorities.
- Is teamwork fostered? Or is it "them vs. us"? In safety, is it "police vs. policed"?
- What is the history? What are the traditions?
- Who are the corporate heroes? And why?
- Is the safety system intended to save lives or to comply with regulations?
- Are supervisors required to do safety tasks daily? This says that safety is a big value.
- Do big bosses wander around? Talk to people?
- Is using the brain allowed on the work floor?
- Has the company downsized?
- Is the company profitable? Too much? Too little?

There are an infinite number of things that set the culture, and we have only listed a few. It is more important to understand what the culture is than to understand why it is that way.

CATASTROPHE CAUSES

We have suggested that culture determines what program elements will work and what elements will not work, and that this culture will dictate final results, what the accident record will be. This is true whether we are looking at frequency or severity. We have evidence today that culture is to a large degree behind human-caused catastrophes. Edwin Zebroski, writing in the book *Risk Management* (a collection of essays put together by GPUN, the organization that runs the Three Mile Island nuclear facility), explains his research into the reasons behind human-error-caused catastrophes:[3]

---

3. Reprinted from Lessons learned from man-made catastrophes by Edwin L. Zebroski, in *Risk Management*, 1991, pp. 51–65, by R. Knief et al., Hemisphere Publishing, Washington, DC. Reproduced with permission. All rights reserved.

When we examine the common factors in large man-made catastrophes, we always find that many relevant and ultimately crucial factors were ignored. Nobody wanted to look at them or put them "on the table" for decision on the protective or remedial actions needed. One of the real benefits of structured decision analysis is to make some of the key factors in a decision *explicit* rather than implicit, and to get in view any "sacred cows" that affect and sometimes put blinders on how decisions are made or delayed.

FOUR MAN-MADE CATASTROPHES

A paper given at the Society of Risk Analysis (SRA) examined four major engineering catastrophes—Three Mile Island, Unit 2 (TMI-2); Chernobyl, Unit 4; the Challenger space shuttle; and the Bhopal chemical plant. These events were selected since each had been investigated thoroughly and impartially, and each of them was well documented. The sequence of events and root causes were well established by literally tens to hundreds of man-years of investigative and analytic efforts. This study sought to find whether there were some common attributes that might help strengthen the urgency of the good practices that the Institute for Nuclear Power Operations (INPO) and the NRC were advocating. And of course, if there were a clear enough set of common attributes, they could be used to identify situations ripe for catastrophes.

Interestingly, this whole approach was triggered by a book by Janis that looked for the attributes common to situations leading to major blunders in U.S. foreign policy. In each case the bad decisions were made by highly respected teams of people with seemingly top qualifications in experience and education. How could such team make decisions which in retrospect were highly suspect on information widely available at the time? (This means *then* current information, and not just in retrospect with "Monday-morning quarterbacking.") The approach used was to look at the technical problems that led to the catastrophes and asking, "what were the decision factors that prepared the situation for the eventual catastrophe?" This approach was surprisingly productive when applied to industrial catastrophes even though the simplifying assumptions were very different from those identified by Janis for his study of foreign policy disasters.

Eleven attributes that were found have had medium to large degrees of commonality in the basis for the TMI-2, Chernobyl, Challenger, and Bhopal events are:

1. *Diffuse responsibilities* with rigid communication channels and large organizational distances from decision-makers to the plant
2. *Mindset* that success is routine with neglect of severe risks that are present
3. *Rule compliance* and the belief that this is enough to assure safety

4. *Team player emphasis* with dissent not allowed even for evident risk
5. *Experience* from other facilities not processed systematically for application of lessons learned
6. *Lessons learned disregarded* and neglect of precautions widely adopted elsewhere
7. *Safety analysis and responses* subordinate to other performance goals in operating priorities
8. *Emergency procedures,* plans, training, and regular drills for severe events lacking
9. *Design and operating features* allowed to persist even though recognized elsewhere as hazardous
10. *Project and risk management techniques* available but not used
11. *Organization* with undefined responsibilities and authorities for recognizing and integrating safety matters

These attributes are generally self-explanatory. A few, however, benefit from some explanation. Safety analysis and risk management can sometimes be subordinate to operating goals, be they for profit, productivity, Deming quality awards, or other purposes. In terms of the lack of emergency procedures for severe events, Bhopal, for example, had weekly drills with gas masks, but what might happen at or beyond the site boundary was not considered. The sirens that were intended to warn the neighborhood were ignored since they were also used every week for the in-plant drills (i.e., they "cried wolf," something NRC drillmasters should note!).

Challenger provides a prime example of design and operating features that were widely recognized and corrected elsewhere but not applied in the particular operation. The earlier Apollo Program had many safety features and applied safety and analysis disciplines that were not used effectively for the shuttle.

Project risk management techniques were readily available, but systematic risk analysis was not applied to any of the four catastrophes. For example, Challenger had 700 "critical items" but with no usable ranking for probability, importance, or interactions. Additionally, the organizational responsibilities for pulling all the safety-related loose ends together were not well defined. The decision makers for launching (under the unprecedented and untested freezing environment conditions) were several levels removed from detailed technical knowledge of crucial hazards present. A crucial comforting statistic, namely the relatively good survival of the booster clevis joints in Titan launches, seemed to overlook the fact that the clevis in question faced *upward* to the rain and ice, while on the Titan it faced *the other way* !

The matrix in Figure 5-1 lists the eleven attributes and identifies the catastrophe to which each was a major contributor. It may be observed that

	Matrix of Common Attributes Four Severe Accidents			
	Bhopal	Challenger	Chernobyl	TMI-2
Responsibilities	X	X	X	X
"Mindset"	X	X	X	X
Rule Compliance	(X)	X	O	X
Team Agreement	X	X	X	X
Prior Events	X	X	X	X
Narrow Experience	X	X	X	X
Output vs Safety	X	X	X	(X)
Severe Accidents	X	X	X	X
Known Hazards	X	X	X	X
Risk Techniques	X	X	X	X
Safety Integration	X	X	X	X

Figure 5-1. Catastrophe vs. attribute matrix. (Reprinted with permission from E. Zebroski, *Risk Management,* 1991, pp. 51–65, Hemisphere Publishing, Washington, DC. All rights reserved.

all of the systems basically share at least ten of the eleven attributes. Chernobyl was surely not excessively dependent on rule-following, since it had very few written rules and none for the test leading to the accident. Bhopal had written rules but [was] chronically out of compliance with many of them. The most solid exception to the eleven attributes was for Three Mile Island, where excessive devotion to production over safety was not a significant contributor to what happened.

The lessons learned from these major accidents are in a sense the mirror images of the good practices. When the Institute of Nuclear Power Operations (INPO) was establishing performance indicators and good practices, one of the concerns was that they not be perceived as just cliches for neatness (i.e., like the old Navy saying, "Sail she might, but shine she will"). In other words, how can "spit and polish" be distinguished from what is really relevant and cost effective to getting the good result? This is a continuing worry, that some performance indicators and good practices may tend to be discounted under the pressures of daily operation. Because they sound like cliches it is sometimes hard to insist that they are always essential. To avoid setting up a situation ripe for a large catastrophe, however, the distribution of attributes in Figure 5-1 should be persuasive that even cliches can be vital.

PIPER ALPHA OIL PLATFORM DISASTER

The previous four accidents have been thoroughly analyzed and are well documented in publications. The emphasis of the paper now shifts to the Piper Alpha oil platform explosion and collapse in July 1988, which re-

Culture Causes 71

cently has become reasonably well documented through the published hearings and studies of the Department of Energy in Great Britain.

Those who have studied the Three Mile Island Accident or the Chernobyl accident, will see an eerie parallelism in the sequence of events for Piper Alpha. Even some of the instrument readings foreshadowing the accident could be overlaid in generic terms on some of those from TMI-2 or Chernobyl. Despite the widely different technologies, similar symptoms were showing up. Piper Alpha is worth considering in more detail since it is another potent example of the degree to which the attributes of catastrophes are definable and readily visible if there is a willingness to look perceptively.

The Piper Alpha oil platform was located in the North Sea 125 miles northeast of Aberdeen, Scotland. It consisted of roughly 20,000 tons of structural steel sitting 400 feet above the sea bed on a 14,000-ton structural-steel "jacket." It had four modules for processing, a boom with a flare, and other booms for derricks and supplies. A large accommodation module included a hotel for over 200 people and a helicopter pad. The platform served 24 regular oil wells plus ten injection wells.

Piper Alpha began operation in 1976. It had experienced a prior small explosion in the compressor area in 1984. A report of that explosion was issued in 1988, arriving *after* the July 1988 accident. This is strangely parallel to the treatment of the incident at the Davis-Besse reactor which occurred 18 months prior and was a direct precursor to the TMI-2 accident. Here also the report did not get to the operators at TMI until after the accident.

On July 6, 1988 Piper Alpha experienced fire, explosions, and structural collapse which led to 167 fatalities (165 on the platform and two in a rescue boat). There were 61 others (almost miraculously) rescued. There was total loss of the platform and wells down to the water line. An extended cleanup project was then required. In most respects, it was a much more serious accident than Three Mile Island, where there were no casualities, but great apprehension in the surrounding communities.

Pre-accident Conditions

Post-accident analyses identified a number of significant conditions that had existed beforehand and contributed to the accident. For one, the Piper Alpha platform was owned by four companies with Occidental Petroleum the operator and, at 36.5%, the largest shareholder. Responsibility on some matters was divided among the four owners.

The platform was licensed in 1972 and built in 1973. This was before the United Kingdom set out and applied enhanced oil-platform construction specifications in 1974. Similarly, the pipelines to other platforms were built just before the Pipelines Act of 1976 became effective. No upgrades to the better specifications were adopted, although there was time to do so. The "Phase 1" type of operation, required when the compressors tripped off line, was unusual. It had not been used since 1984. There were no written procedures available, nor had there been training for this condition.

The Plant Superintendent was away, so the operating superintendent, deputy, and lead operator each had been promoted one level temporarily. At the same time, a "minimum production technique" was being used with a "minimum production team" of five. The gas operator, the key position in the gas producing area where the accident initiated, had just arrived on the platform and was in his first day on the job!

The safety valve on the "A" compressor was removed for maintenance and blind flanges were installed, but with no quality assurance records or tag-out indications in the control room. The piston compressors had been found to have 30 to 40% of their head studs broken in February–May of 1988. The broken studs were replaced but the non-broken studs were not inspected or replaced. As noted previously, there had been a prior explosion in the compressor area 24 March 1984 for which the report was withheld until October 1988. Other probable accident contributors include:

- Power supplies and cables were vulnerable to small fires
- Isolation valves for the safety valve could be operated from the control board (without indication of a blind flange)
- Key gas alarms were on the back of the control board and were not checked regularly
- Shift-change briefing did not provide awareness of safety-valve maintenance
- Fire walls were nominally, but not adequately, designed to withstand possible open-air gas explosions (at > 4.5 psi).

Conclusions

Compared to the four other catastrophes, ten of the eleven attributes leading to accidents (e.g., Figure 5-1) were evident for Piper Alpha. The eleventh— training and procedures—could be said to be effective for normal operation, but was neglected for "Phase 1." Overall, a high degree of likelihood of a severe accident was foreseeable in terms of the attributes of the previously mentioned catastrophes.

LESSONS FROM CATASTROPHES IN ENGINEERING OPERATIONS

The conclusion from discussions of Piper Alpha is that catastrophes in engineering operations (or at least the five discussed here), share all or most of the 11 attributes listed. Theoretical exceptions, involving external natural

Culture Causes

catastrophes or wars, may be postulated as producing disasters even for organizations that do not have these negative attributes. However, even for such cases it seems clear that organizations that have the positive attributes that are the mirror images of the negative ones would survive better, salvage more, and recover much sooner than organizations that have these negative attributes.

One important generalization from such catastrophes is that *major engineering catastrophes are rarely if ever accidental!* They are not accidental in the sense of being highly foreseeable and predictable if the conditions described in the attributes are present. Historically, large engineering catastrophes have often been written off as unforeseeable "acts of God." The initiating events sometimes may indeed be portrayed to be bizarrely unlikely, and beyond the most prudent management's ability to foresee. However, the lesson of the catastrophes studied so far is that bizarre (that is, unexpected, very infrequent, or unprecedented) events are not particularly rare or unforseeable if one looks at the experience derived from a large population of engineering operations. The situation that sets the stage for an initiating event to mushroom to a catastrophe is definable in terms of the eleven attributes of Figure 5-1. The converse of the accident attributes constitutes a basic list of "good practices." With good practices, even the bizarre initiating events cause only localized damage and the operation overall can remain healthy, or recover gracefully.

ASSESSING CULTURE

While you already have a sense of the culture if you have been with an organization a while, to get a truly accurate picture you must question the working employees—the hourly people.

In the book *Managing Employee Stress,* we suggested ways to assess organizational culture as a primary cause of job stress:[4]

> The first technique is to get information on employee health disturbances (emotional, behavioral and health deterioration) and determine the trends of the employee population. One way of doing this is through the use of questionnaires asking anonymously about:
>
> - Tenseness on the job
> - Anxiety

---

4. Reprinted from D. Petersen, *Managing Employee Stress,* Aloray, Inc., Goshen, NY, 1990.

- Sleeping problems
- Depression
- Fatigue
- Inability to concentrate
- Aches and pains
- Headaches
- Satisfaction with job
- Alcohol consumption
- Increased smoking
- Drug abuse

Information from other sources will tell you about the trends in:

- Accidents
- Absenteeism
- Turnover

Information may (or may not) be available on trends in stress-related diseases of the work force.

Questionnaires might also be used to ask anonymously about such items as:

- Job control
- Importance of work
- Relationships with the boss
- Pressures from workload, time

Another information source is to look at such things as:

- Repetitiveness of work
- Machine speeds
- Forced pace
- Forced overtime
- Downsizing
- Isolation
- Need for vigilance
- Demands for preciseness
- People being rushed
- Seriousness of mistakes
- Unsafe behavior

You might look at what support systems are available to the workers within the company or in the area. In-depth interviews with hourly employees will often unearth much of what you need to know to assess the real situation.

Traditional safety measures do not tell us much. Numbers of accidents

Culture Causes

may or may not be an indicator. Accidents correlate very highly with stress-producing environments, but it takes an extremely large unit to generate enough accidents to attain statistical validity. In most cases they become relatively meaningless figures. Audits give little indication as they typically tell us little about the existing climate.

Interviews and surveys, however, attempt to find out from all levels of the organization what climate really does exist. If properly done, they are valid and very meaningful.

Interviews

Asking employees can be very useful in assessing the real corporate climate, assuming they will level with you. Whether or not this happens is probably dependent on trust and spending enough time in the process to get to where they will level with you.

Often it takes an outsider to get to the truth from the employees. This fact in and of itself often suggests a problem exists. It also usually takes time to get to reality in interviews. I have found that in most organizations it takes 10–15 minutes of general talk during an hourly interview before anything important begins to emerge. It takes that long for trust to be established. Once established it often takes an hour or more to close off the out-pouring of facts and feelings that emerge. When this relationship is established, the interview is probably the best technique to get to reality. But do not expect good data from short interviews.

Surveys

The second option is the survey. With this approach a lot of data can be gathered quickly and it can be processed rapidly by computer. But the survey instrument is crucial. It must be constructed by people who know what they are doing and that takes time. Each question must be carefully constructed and validated by a professional test development person working with subject experts. We have only recently been blessed with some general safety climate surveys (and only a few are currently on the market).

One of the surveys, constructed in a ten-year research project of the Association of American Railroads (AAR), could provide good insight, I believe. This survey looks at 20 categories of safety system success as perceived by each level of the organization and since many categories do relate to organizational causes of stress, it could be useful. [See Figure 5-2.]

The important thing in this survey is that the survey scores correlate very highly with bottom line safety results.

This survey is one good indicator. Others are on or will be coming on the market. You cannot assure these will be useful, however, without information on construction and validation of the instrument.

Organizational variables	SYSTEM 1	SYSTEM 2	SYSTEM 3	SYSTEM 4	Item No.
LEADERSHIP					
How much confidence and trust is shown in subordinates?	Virtually none	Some	Substantial amount	A great deal	1
How free do they feel to talk to superiors about job?	Not very free	Somewhat free	Quite free	Very free	2
How often are subordinate's ideas sought and used constructively?	Seldom	Sometimes	Often	Very frequently	3
MOTIVATION					
Is predominant use made of 1 fear, 2 threats, 3 punishment, 4 rewards, 5 involvement?	1, 2, 3, occasionally 4	4, some 3	4, some 3 and 5	5, 4, based on group	4
Where is responsibility felt for achieving organization's goals?	Mostly at top	Top and middle	Fairly general	At all levels	5
How much cooperative teamwork exists?	Very little	Relatively little	Moderate amount	Great deal	6
COMMUNICATION					
What is the usual direction of information flow?	Downward	Mostly downward	Down and up	Down, up, and sideways	7
How is downward communication accepted?	With suspicion	Possibly with suspicion	With caution	With a receptive mind	8
How accurate is upward communication?	Usually inaccurate	Often inaccurate	Often accurate	Almost always accurate	9
How well do superiors know problems faced by subordinates?	Not very well	Rather	Quite well	Very well	10

Organizational variables	SYSTEM 1	SYSTEM 2 Policy at top, some delegation	SYSTEM 3 Broad policy at top, more delegation	SYSTEM 4 Throughout but well integrated	Item No.
DECISIONS					
At what level are decisions made?	Mostly at top				11
Are subordinates involved in decisions related to their work?	Almost never	Occasionally consulted	Generally consulted	Fully involved	12
GOALS					
What does decision-making process contribute to motivation?	Not very much	Relatively little	Some contribution	Substantial contribution	13
How are organizational goals established?	Orders issued	Orders, some comments invited	After discussion, by orders	By group action (except in crisis)	14
How much covert resistance to goals is present?	Strong resistance	Moderate resistance	Some resistance at times	Little or none	15
CONTROL					
How concentrated are review and control functions?	Very highly at top	Quite highly at top	Moderate delegation to lower levels	Widely shared	16
Is there an internal organization resisting the formal one?	Yes	Usually	Sometimes	No—same goals as formal	17
What are cost, productivity, and other control data used for?	Policing, punishment	Reward and punishment	Reward some self-guidance	Self-guidance problem-solving	18

1 2 3 4 5 6 7 8 9 10 11 12 13 14 15 16 17 18 19 20

Figure 5-2. Organizational profile. (From D. Petersen, *Managing Employee Stress*, Aloray, Inc., Goshen, NY, 1990.)

Another indicator is general morale surveys. Many are available. One of the best can be found in the book, *The Human Organization,* by Likert.

Other Instruments

Other instruments that are available are included in the Appendix. Any you use must be interpreted, and this may require professional help.

Interpreting the Instruments

Correct interpretation is, of course, crucial. For instance, in using the above mentioned AAR Safety Climate survey, one organization generated the

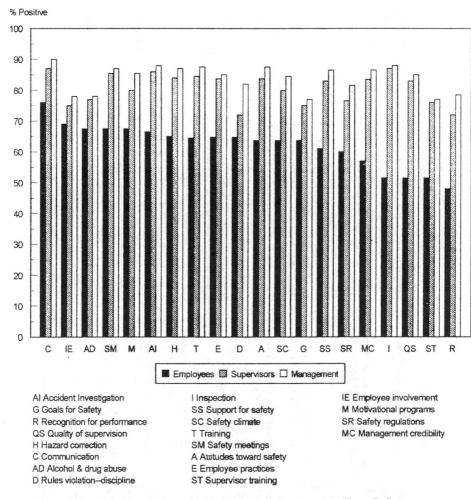

AI Accident Investigation
G Goals for Safety
R Recognition for performance
QS Quality of supervision
H Hazard correction
C Communication
AD Alcohol & drug abuse
D Rules violation—discipline

I Inspection
SS Support for safety
SC Safety climate
T Training
SM Safety meetings
A Attitudes toward safety
E Employee practices
ST Supervisor training

IE Employee involvement
M Motivational programs
SR Safety regulations
MC Management credibility

Figure 5-3. Safety survey response (railroad—all departments). (From D. Petersen, *Managing Employee Stress,* Aloray, Inc., Goshen, NY, 1990.)

Culture Causes

	What is the level of job satisfaction?	Turnover?
What is the size of your community?	What are the physical conditions?	Absenteeism?
How much publicity has there been on stress in your company, your community?	Are workloads a problem?	More accidents?
	Are work hours a problem?	Low creativity?
Merger?	Are roles clear? Blocked career paths?	Accept mediocrity?
Lay-offs?	Adversial relations?	Is work routine?
Are your people overloaded?	What management style predominates?	Lack of participation?
Is there boredom on the job?	Takeover possibilities?	
	Are you downsizing? Loss of felt loyalty? Enough training? Politics a problem? High automation? Use piece rates? Is machine pacing a problem? Shift work a problem? Forced overtime?	

Figure 5-4. Organizational factors to consider. (From D. Petersen, *Managing Employee Stress,* Aloray, Inc., Goshen, NY, 1990.)

chart in Figure 5-3. What does it all mean? A quick analysis of the chart shows the worst results came in certain categories: recognition, quality of supervision, supervisory training and inspections, all indicators of a lack of supervisory performance. A relatively high score was obtained on employee involvement. People want to contribute, but their bosses don't really know how to make it happen. This could result in a deep frustration.

Surveys like the above give clear indication of problems and provide a diagnosis as to what ought to be done.

These are some of the ways we might assess the safety climate of the organization. Figure 5-4 summarizes these measures and adds a few more.

REFERENCES

Bailey, C. *Using Behavioral Techniques to Improve Safety Program Effectiveness.* Washington, DC: Association of American Railroads, 1988.

Likert, R. *The Human Organization,* New York: McGraw-Hill, 1967.

Minor, J. and J. Brewer. The management of ineffective performance. In *Handbook of Industrial and Organizational Psychology,* edited by M. Dunnette. Chicago: Rand-McNally, 1976.

Peters, T. and R. Waterman, *In Search of Excellence,* New York: Harper & Row, 1982.

Petersen, D. *Managing Employee Stress,* Goshen, NY: Aloray, Inc., 1990.

Zebroski, E., Lessons learned from man-made catastrophes, in *Risk Management,* by R. Knief et al., New York: Hemisphere Publishing, 1991.

Chapter 6

Design Causes

This chapter discusses design effects on humans from two quite different perspectives:

1. How design causes human error.
2. How design causes cumulative trauma disorders.

Both of these are an integral part of the general field of ergonomics, but we will separate them because of the strong emphasis on the second issue at the time of this writing, to the almost total lack of regard for the first, which represents a much bigger cause of injury and error.

Using our model(s) from Chapter 3, this chapter deals with traps; the traps that management unknowingly sets for employees. Since traps are a way to describe human factor concepts, this chapter deals with the discipline we know as *human factors engineering,* a discipline that by its very definition is safety-related. While human factors engineering is not an exact science, it is a rational approach to the problems of designing and constructing things used in the workplace so that the users will be less likely to make errors resulting in accidents. It attempts to make machines more convenient and comfortable, less confusing, less exasperating, and less fatiguing.

Human factors engineering is also called ergonomics. Elsewhere it is called *biomechanics, biotechnology, psychophysics, biophysics, human engineering, human factors,* and *engineering psychology.*

Here are three simple definitions:

1. It is engineering for the population that will use it.
2. It is designing the system so that machines, human tasks, and the environment are compatible with the capabilities and limitations of people.

3. It is designing the system to fit the characteristics of people than retrofitting people to suit the system.

Ergonomics is interdisciplinary in nature, as is safety. Figure 6-1 shows its roots and relationships. The figure is by Francis Dukes-Dubos, who describes the field as follows:[1]

> The term ergonomics has its origin in the Greek: ergos = work, nomikos = law. The term was adopted in England in 1949 to label the science that deals with the many-sided problems of how to fit a job to man's anatomical, physiological, and psychological characteristics to enhance human efficiency and well-being.
>
> It is quite widely held that human engineering and human factors engineering are synonymous with ergonomics—the former terms being the preferred ones in the United States, whereas the latter is more favored in Europe.
>
> The classic definition of human engineering as formulated by Hatch suggests that this discipline, in accordance with its name, deals primarily with the engineering aspects of ergonomics—that is, the application of laws governing the man–machine system to the analysis and design of machines and other structures of the work environment.
>
> Human factors engineering is taught and practiced mostly by psychologists; thus the major emphasis on the behavioral aspects of the man–machine system. Ergonomics is broader because it deals not only with application of the laws but also with establishing the laws governing the man–machine system.

Ross McFarland of Harvard was one of this country's leading human factors theorists and did much to apply the theories to the field of safety. He traced the history of the discipline in the United States as follows:[2]

> Human factors as a specialized discipline developed largely as a result of the use of complicated equipment during World War II. Modern aircraft and weapons systems did not place demands on an operator's muscle power but upon his sensory, perceptual, judgmental and decision-making abilities. The potentialities of numerous complex, precisely-engineered devices could not be realized because their operational requirements exceeded the capabilities of their operators. The solutions often depended on the devel-

---

1. Reprinted with permission from F. Dukes-Dubos, The place of ergonomics in science and industry, *ASSE Journal*, October 1972.
2. Reprinted from R. McFarland, Applications of human factors engineering to safety engineering problems, in J. Widener, Ed., *Selected Readings in Safety*, Macon, GA, Academy Press, 1973.

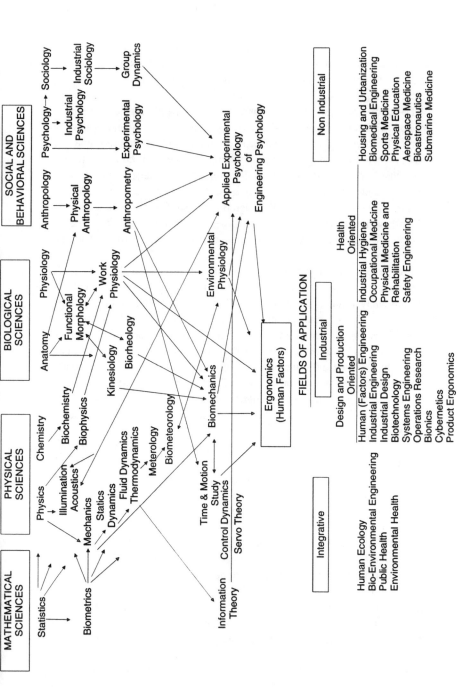

Figure 6-1. Schematic illustration of the interdisciplinary nature of human factors engineering (ergonomics). (From F. Dukes-Dubos, The place of ergonomics in science and industry, *ASSE Journal*, October 1972.)

opment and application of data from such fields as experimental psychology, anthropology, physiology, biology and medicine, working in collaboration with engineers. Only in this way was it possible to improve the integration of men and the new complex machine systems.

The efforts of these scientists, largely working within the aircraft industry and in military research and development, resulted in the emergence in the USA of human factors as a special discipline. Initially, those trained in experimental psychology played a predominant role, and the early studies were chiefly concerned with design features of dials, gauges, controls, and consoles, and in many specialized areas such as speech communication in the presence of aircraft noise. Although there was relatively greater emphasis on psychology in the early development, the contributions from other disciplines in the biological and behavioural sciences also were of major importance.

Since World War II the growth of human factors in America has been very rapid. It has been extended through the technological developments in systems analysis and in computer capabilities for simulation and modelling. The combined disciplines comprising the field have had their most extensive applications in the design of space vehicles and in space flight, as well as in nuclear submarines and other undersea operations. Increased application of the methods and principles have also occurred in more prosaic areas, such as in the design, not only of air and ground transport equipment, but also of telephones, farm tractors, typewriters, kitchen stoves, housing for the elderly, and child-proof containers for medicines and toxic household agents.

Prior to 1950 only one industrial organization had established a formal human factors programme in America. By 1956, however, this number had grown to 24 companies, with an average of about 20 persons per group, and by 1968 there were formal programmes in about 175 companies. Numerous other human factors specialists have been employed in university and non-profit research centres and in government agencies.

In spite of these favourable developments, it is unfortunate that the contributions of these disciplines are often ignored in solving many of the problems which confront us. Thus far, only limited applications of human factors principles have been made to the field of industrial design. One can go into any factory and find many instances where design factors relating to worker characteristics have been ignored. Concrete evidence is available from the records of injuries, first-aid treatments, work spoilage or machine malfunction.

To use the knowledge available to us from the field of human factors, we might consider first what we need to know about people's capabilities and expectations, and then begin to consider how we should use this information in designing the work environment.

Design Causes

As indicated above, we will lightly touch on ergonomics on this chapter from the standpoints of how design causes human error and how design causes cumulative trauma disorders.

HOW DESIGN CAUSES ERROR

Most of the theorists and writers in the fields that look at human error tend to come from rather technical backgrounds. While some have psychology and management backgrounds, as mentioned earlier, most come from the technical areas of systems engineering, reliability engineering, systems safety, and human factors engineering. Consequently, much of the writing about human error is technical and involves systems concepts. A number of investigators have discussed the human being in this context, and most put primary emphasis on an examination of the worker–machine interface and relationship. Typical is the writing of Dr. Ross McFarland, who developed the model shown in Figure 6-2.

McFarland looked at the relationship between worker and machine as a primary cause of human error. On the worker side of this relationship, he described a human being's sensory organs as receiving information from the machine's displays, then transmitting information to the central nervous system (all this is influenced by the

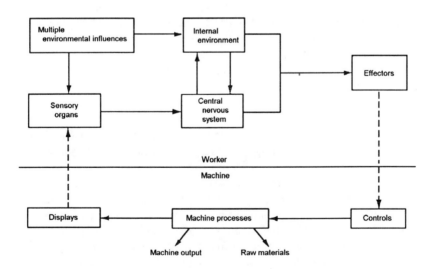

Figure 6-2. The interacting components of a typical worker–machine system. (From R. McFarland. Applications of human factors engineering to safety engineering problems, in J. Widener, Ed., *Selected Readings in Safety*. Macon, GA: Academy Press, 1973.)

worker's internal and external environments, and then the worker's effectors manipulating the machine's controls to perform the machine processes that take raw materials and convert them into some product or part.

A similar model is provided by Wood (Figure 6-3), who examines the human part of the worker–machine system in more detail. He describes the human mind as consisting of four components: (1) an input analyzer, (2) feelings and thoughts, (3) an analysis and decision-making process, and (4) a behavior identifier. These four components of the human mind interact regularly with both external inputs and internal inputs, as shown, and attempt to control the situation using some output. The output is the actual behavior of the individual, which might be quite different from his or her expected behavior (a behavioral error).

The Worker's Sensing Subsystem

One of the worker's functions is that of a sensor, an information seeker. And human beings have more than the five senses; we have many sensors. To use the various senses in the work environment, we build information displays—devices that gather needed information and translate that information into inputs the human brain can perceive. These displays are in two general classes: pictorial and symbolic.

Symbolic displays present the information in a form that has no resemblance to what it is measuring. Some examples are the speedometer, thermometer, pressure gauge, and altimeter. In pictorial displays, the geometric and spatial relationships are shown as they exist. Maps, pictures, and television are examples of pictorial displays.

The two most common types of symbolic displays are visual and auditory. Much attention has been given to the design of these types of displays, and some general principles have emerged:

1. The principle of simplicity: The purpose for which the display is to be read will dictate its design, but as a general principle the simplest design is the best.
2. The principle of compatibility: The principle of compatibility holds that the motion of the display should be compatible with (or in the same direction as) the motion of the machine and its control mechanism.
3. The principle of arrangement: As the design of the display is important, so too is its location or arrangement in relation to other displays. A poor arrangement of displays can be the source of error. Sometimes dials must be arranged in groups on a large control panel. If all the dials must be read at the same time, they should be pointing in the same direction within the desired range. This will reduce check-reading time and increase accuracy.

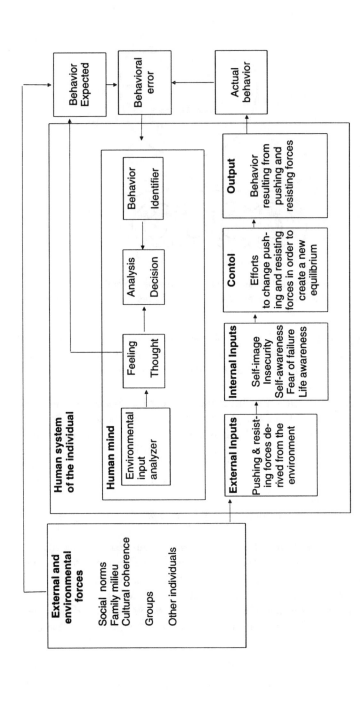

Figure 6-3. Systems model for the analysis of behavioral change. (From F. Wood, in Gordon L. Lippitt, *Visualizing Change*. La Jolla, CA: University Associates, 1976. Reprinted with permission of The Gordon Lippitt Foundation, Bethesda, MD. All rights reserved.)

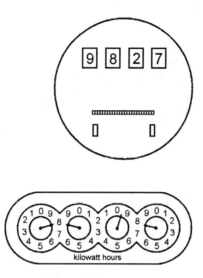

Figure 6-4. Electric meters. The one on top illustrates the principle of simplicity. (From F. Vilardo, Human factors engineering, *National Safety News,* August 1968.)

4. The principle of coding: All displays should be coded or labeled so that the operator can tell immediately to what mechanism the display refers, what units are measured, and what the critical range is.

Some of these principles are illustrated fairly easily. Figure 6-4 illustrates the principle of simplicity. Of the two electric meters shown, the top one is considerably easier to read and much less likely to cause a sensing error. In Figure 6-5, the drawing on the right illustrates the principle of compatibility. To regulate heat, the user simply turns the pointer clockwise to "Hi." The principle of arrangement is shown in Figure 6-6. The configuration on the left is better than the one on the right and is much less likely to cause errors.

With regard to auditory displays, Frank Vilardo provides these principles:[3]

Situationality: The design of auditory displays should consider other relevant characteristics of the environment in which the system is to function (e.g., noise levels, types of responses controlled by the auditory signal).

Compatibility: Where feasible, signals should "explain" and exploit learned or natural relationships on the part of the users, such as high frequencies being associated with "up" or "high" and wailing signals indicating emergency.

---

3. Reprinted with permission from F. Vilardo, Human factors engineering, *National Safety News,* August 1968.

Design Causes

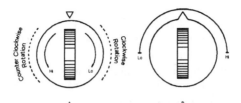

Figure 6-5. Heat regulators. The one on the right illustrates the principle of compatibility. (From F. Vilardo, Human factors engineering, *National Safety News,* August 1968.)

Approximation: Two-stage signals should be considered when complex information is to be displayed and a verbal signal is not feasible. The two stages should consist of a) attention-demanding signals to attract attention and identify a general category of information and b) designation signals to follow the attention-demanding signals to designate the precise information within the general category.

Dissociability: Auditory signals should be easily discernible from other sounds (be they meaningful or noise).

Parsimony: Input signals to an operator should not provide more information than is necessary to carry out the proper response.

Forced entry: When more than one kind of information is to be presented, the signal must prevent the receiver from listening to just one aspect of the total signal.

Invariance: The same signal should designate the same information at all times.

The Worker's Information-Processing Subsystem

The information-processing system involves these kinds of processes:

Information storage, including both long- and short-term memory.
Information retrieval and processing, including:

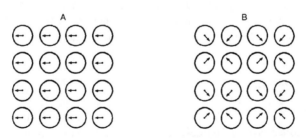

Figure 6-6. Indicators. The grouping on the left illustrates the principle of arrangement. (From F. Vilardo, Human factors engineering, *National Safety News,* August 1968.)

- Recognition or detection of signals or stimuli.
- Recall, including recalling both previously learned factual information, and information in short-term storage.
- Information processing, categorizing, calculating, coding, computing, and so on.
- Problem solving and decision making.
- Controlling of physical responses.

Each of these is a highly complex area and is a major field of study in itself, beyond our scope. Suffice it to say that it would be easy to overload this subsystem in certain difficult or stressful situations. Probably, in most industrial settings, this is less a problem than are some of the other subsystem overloads.

The Worker's Responding Subsystem

The third function the worker serves in the worker–machine system is that of a controller. This control function is the response that the worker must make to any given stimulus. Just as principles exist for designing displays less likely to cause errors, they exist for designing controls:

1. The principle of compatibility: Control movement should be designed to be compatible with the display and machine movement.
2. The principle of arrangement: This principle provides for the grouping of elements or components according to their functions (those having related functions are groups together) and for grouping in terms of how critical they are in carrying out a set of operations. It also suggests that each item be in its optimum location in terms of some criterion of usage (convenience, accuracy, speed, strength to be applied, etc.). Items used in sequence should be placed close together, and items most frrequently used should be placed closer to the operator than items less frequently used.
3. The principle of coding: Whenever possible, all controls should be coded in some way. A good coding system uses shape, texture, location, color, and operation. Also, all controls and displays should be labeled. Labeling is crucial if the operators change often or the equipment is shared.

Human Expectations

Any situation that calls for a movement contrary to the established stereotype is bound to produce errors. The designer is calling for errors (building traps) by asking the worker to change, in this or that unique situation, a behavior pattern that can be described as a habit. People expect things to be, or work, in a certain

Design Causes

manner. The following are some human stereotypes that should be considered in the design of both displays and controls:

- Handles used for controlling liquids are expected to turn clockwise for off and counterclockwise for on.
- Knobs on electrical equipment are expected to turn clockwise for on or to increase current and counterclockwise for off or decrease in current. (*Note:* This is opposite to the stereotype for liquids.)
- Certain colors are associated with traffic, operation of vehicles, and safety.
- For control of vehicles in which the operator is riding, the operator expects a control motion to the right or clockwise to result in a similar motion of the vehicle, and vice versa.
- Sky–earth impressions carry over into colors and shadings: light shades and bluish colors are related to the sky or up, whereas dark shades and greenish or brownish colors are related to the ground or down.
- Things that are farther away are expected to look smaller.
- Coolness is associated with blue and blue-green colors, warmness with yellows and reds.
- Very loud sounds or sounds repeated in rapid succession, and visual displays that move rapidly or are very bright, imply urgency and excitement.
- Very large objects or dark objects imply "heaviness." Small objects or light-colored ones appear light in weight. Large, heavy objects are expected to be "at the bottom." Small, light objects are expected to be "at the top."
- People expect normal speech sounds to be in front of them and at approximately head height.
- Seat heights are expected to be a certain level when a person sits down.

For additional design data see Appendix C.

How Design Causes Cumulative Trauma Disorders

The second part of ergonomics, which is highly emphasized today, is concerned with how design causes cumulative trauma disorders. While not a common problem in the earlier years of safety, this issue now receives widespread attention.

How serious a problem this really is, is subject to some debate. OSHA, in its early editions of ergonomic guidelines, stated that it constituted one-third of our accident problem. Although this may be true if we include back claims in the total, it surely does not apply to the injuries we usually associate with repetitive motion, that is, soft tissue injures of the wrist, shoulder, and neck. It is difficult to define the extent of this problem. State of California statistics show that repetitive motion injuries account for only 0.7% of the worker's compensation claims in that state, a far cry from the OSHA figures. In short, we simply do not know how serious the problem is.

Still, major attention is being paid to this subject today, rightly or wrongly. Because of labor disputes in the meat packing industry followed by massive media publicity and several multi-million-dollar fines, we now spend a great deal of time on cumulative trauma disorders.

As the emphasis in this book is on system-caused human error, rather than system-caused illness, we spend little time on this issue, but we have included relevant informtion in Appendix B, which looks at an abridged version of OSHA's early attempts at an ergonomic guideline for all industry, based on a previously published guideline for the meat packing industry. That guideline is fairly comprehensive (some 80+ pages) and covers virtually everything that a company could do in this area. Whether or not doing all this would be practical to prevent 0.7% of the losses (California figures) is open to some discussion.

For instance, in this early attempt, the guidelines propose:

- A policy placing safety and health on the same level of importance as production.
- That each manager, supervisor, and employee be held accountable for carrying out specified things.
- That there be a complaint procedure.
- A procedure for each employee to report any symptom of a CTD (a pain).
- That there be a written program with:
  - Analysis of trends
  - Employee surveys
  - Plant evaluations
  - Logs of problems, solutions, and improvements
  - Review
  - Work site analysis
  - Analysis of all medical records
  - Employee questionnaries regularly
  - Checklists and video tape analyses of all jobs
  - An ergonomist to oversee
  - Job rotation
  - More rest breaks
  - Additional people
  - Lowered line speeds
  - Job design to the person currently doing the job
  - New employee conditions periods
  - Limit on overtime
  - Lowered production rates
  - Standby/relief personnel

- An Occupational Health Nurse (OHN) to supervise the program
- A regular symptoms survey
- Training for all employees
- Completion of six checklists and two surveys
- Surveys from employees directly to OSHA

While this is no doubt good ergonomics, it is doubtful that industry will do all of this.

REFERENCES

Dukes-Dubos, F. The place of ergonomics in science and industry. *ASSE Journal,* October 1972.

McFarland, R. Ergonomics around the world: the United States of America. *Applied Ergonomnics* 21:19–25, 1971.

Vilardo, F. Human factors engineering: what research found for safety men in Pandora's box. *National Safety News,* August 1968.

Wood, F. In Lippitt, G. *Visualizing Change,* La Jolla, CA: University Associates, 1976.

Chapter 7

Reducing System-Caused Error

The last three chapters have discussed system-caused human errors: errors caused by management, by designs, and by culture. This chapter provides some ideas on control in the three categories.

REDUCING MANAGEMENT-CAUSED ERROR

To control management-caused errors, we must start an assessment of the system—what currently exists. As indicated in Chapter 4, there are a variety of ways to do this, ranging from the simple checklists to the more sophisticated fault trees and profiling systems of that chapter.

Using these techniques will tell you what exists in the organization, which, as indicated earlier is the first of a three-step process:

1. Determine what exists.
2. Determine what should exist.
3. Provide the difference.

The audits and checklists of Chapter 4 are of little help in determining what should exist, however; they only offer another person's opinion on what should be in your organization's "safety program." Actually these audits are even limited in telling you what does exist, for they often reflect only what some manager thinks is going on. This can be far from reality.

A better alternative is always to ask the people who really know, the workers, who can tell you both (1) what does exist and (2) what needs to exist to get results.

The perception survey discussed in Chapter 5 is one way to get the needed information. Another way is through in-depth worker interviews, but that usually is more costly and time-consuming.

Figure 5-3 shows how a perception survey (asking the workers) can answer these two questions. This survey, from a large multilocation organization, clearly illustrates what is strong and what is weak in that organization's safety system, and suggests what is needed for results:

1. The employees believe that there is a serious need for more recognition for the work they do, and management does not know it.
2. The workers believe strongly that there is a supervisory performance problem.
3. There is a credibility problem for management.

Even what should exist in the safety system is highlighted:

1. There needs to be a systematic way to ensure that positive reinforcement occurs.
2. Supervisors need to be held accountable for defined safety performances.
3. Management needs to visibly demonstrate safety's priority.

Reducing Culture-Caused Error

Reducing culture-caused errors requires the same process as reducing management caused-errors:

1. Determine where you are.
2. Determine where you need to be.
3. Provide the difference.

And the process is similar—asking the people who know that the culture is, the workers. A perception survey is one of the best tools we have.

In addition, many other things influence culture, as indicated in Figure 5-4: past or impending mergers, downsizing, boring work, past leadership styles, manager turnover, lack of hourly involvement in the past, being machine-paced, overtime, fuzzy roles, fuzzy goals, and many others. Although many of these factors cannot be changed, their effects must be recognized and managed.

Of crucial importance is an assessment of how workers are measured and rewarded, which is a major determinant of workers' perception of culture. Safety cannot be perceived as a key corporate value if it is seldom or fuzzily rewarded and measured.

REDUCING DESIGN-CAUSED ERROR

Several people have written on this topic and provided some help. Johnson, in MORT, provided a fault tree on the elimination of design-caused errors (Figure 7-1).

Also Woodson developed a model on the prediction and control of human error (Figure 7-2). In the Woodson model, error sources are identified, the probability of their occurrence is estimated, the criticality of their effects is estimated, trade-offs are considered to control the situation, and controls are incorporated into the system design.

Both models speak to the general area of using ergonomic concepts in design to reduce error.

One basic approach to industrial safety relates the physical characteristics and capabilities of the worker to the design of his or her equipment and to the layout

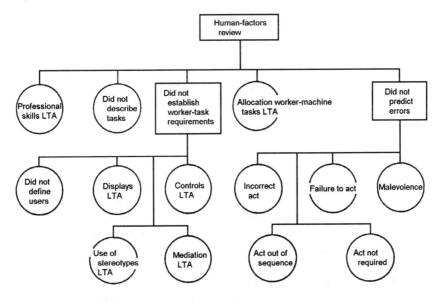

LTA = less than adequate.

Figure 7-1. Human-factors fault tree. (Adapted from W. G. Johnson, *The Management Oversight and Risk Tree—MORT*. Washington, DC: U.S. Government Printing Office, 1973.)

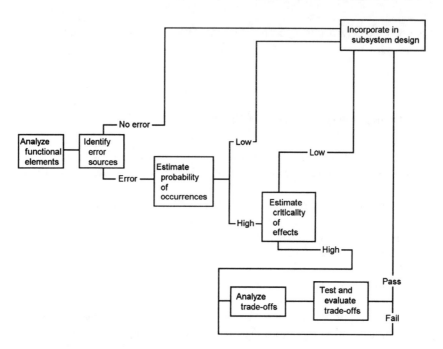

Figure 7-2. Basic analytical flow diagram showing human error, prediction, and control sequence. (From W. Woodson and D. Conover, *Human Engineering Guide*. Berkeley: University of California Press, 1966.)

of the workplace. In order to carry out this approach, several different types of information are needed, namely, a description of the job, an understanding of the kinds of equipment that will be used, a description of the kinds of people who will use the equipment, and, finally, the biological characteristics of these people. In general, the first three items (job, equipment, and users) can be defined both accurately and easily. The worker's biological characteristics are, however, often overlooked or ignored in our own design.

The anthropometric, or human-body-size, data needed consist of various height, length, and breadth measurements that are used to establish the minimum clearances and spatial accommodations and to allow for arm, leg, and body movements made by the worker during the performance of his or her task. We also need data relating to the range, strength, speed, and accuracy of human body movement. Knowledge of human strength in single maximum exertions and over prolonged periods of time is important for the design of equipment to ensure maximum safety in operation. The specific kinds of strength data needed vary from task to task, but might include the maximum weights that can be lifted or carried, or the maximum forces that can be exerted in operating a wheel, lever, or knob during single applications and over extended periods.

ERGONOMIC SYSTEMS ANALYSIS

The *Human Engineering Guide to Equipment Design* by Van Cott and Kinkade provides some directions for human-factors analysis of organizations:[1]

> System analysis, also sometimes called system engineering, can be described as having the following general purposes:
>
> 1. Scheduling. In the development of a complex system, system analysis is necessary to identify all of the requirements and the logical and sequential order in which they must be accomplished.
> 2. Identifying Limiting Factors. System analysis enables the designer to determine the factors that potentially limit or constrain the performance of a system.
> 3. Establishing System Performance Criteria. System analysis provides the criteria which must be met by the several interrelated functional elements of a system. These criteria thus become standards both for design as well as for test and evaluation.
> 4. Identifying and Explicating Design Operations. Through explicit comparison of expected performance with criteria, system analysis enables the designer to decide better the utilization of men and machines.
> 5. Evaluation of Systems. System analysis is a prerequisite to the test and evaluation of systems. Performance measures of the system and its subsystems are needed to find out how a system can be expected to perform under actual operating conditions or whether one system can be expected to be better than another.
>
> System analysis includes the following steps or phases:
>
> 1. The explication of system requirements and constraints.
> 2. The description of system functions.
> 3. Detailed descriptions of operational event sequences (including environmental conditions).
> 4. Detailed descriptions of component processes.
>
> While these steps are interdependent, they are not always followed in a rigid sequence. Some parts of the total system may already have been designed and tested; and time constraints may change the order of approach. Nevertheless, they form a conceptual structure helpful to under-

---

1. Reprinted from H. Van Cott and R. Kinkade, *Human Engineering Guide to Equipment Design*, Washington, DC, U.S. Government Printing Office, 1972.

standing the relationships of human engineering to the total system engineering process.

Figure 7-3 provides a broad outline of some of the things to be considered in making an ergonomic systems analysis.

Although an ergonomic systems analysis begins the process of examining the entire system (like an audit), ergonomics also uses a micro-look, such as JSAs, to probe more deeply. This approach is called task analysis.

Task Analysis

The objective or purpose of task analysis is to provide the basic "building blocks" for the rest of the human engineering analysis. The development of this information through task analysis can be divided sequentially into two major parts: (1) subtask derivation and (2) skill and knowledge analysis. In subtask derivation, information pertinent to the entire subtask is obtained, including the location at which it is performed, and its relationship to existing tasks. To find out which skills and knowledge are required involves an examination of the various steps or parts of the subtasks. This analysis results in a statement of the psychological requirements of the tasks, the kinds of discriminations that must be made, the decision-making, and motor and other skilled responses required.

The kinds of information that subtask derivation may include are:

1. A statement of the task, as derived from the personnel function. The statement should contain an action verb and indicate the purpose of the task in terms of a system goal or subgoal; an example would be the words "actuate power switch."
2. The category of task should be noted, i.e., whether it is an operator, maintenance, or support task.
3. The location in which the task is performed. This separates tasks performed in such separate places as the maintenance shop, supply area, manufacturer's plant, on board ship, in flight, on the flight line, etc. One possible format for the analysis is shown in Table 7-1.

Because specific error information is difficult to obtain during the development stage, it may be difficult or impractical to require specific estimates of the probability of error occurrences. The following general scale may be used, however, to supplement the categorization and description of the error.

1. Error *highly likely* to occur unless specific avoidance technique is utilized (probability greater than 0.5).

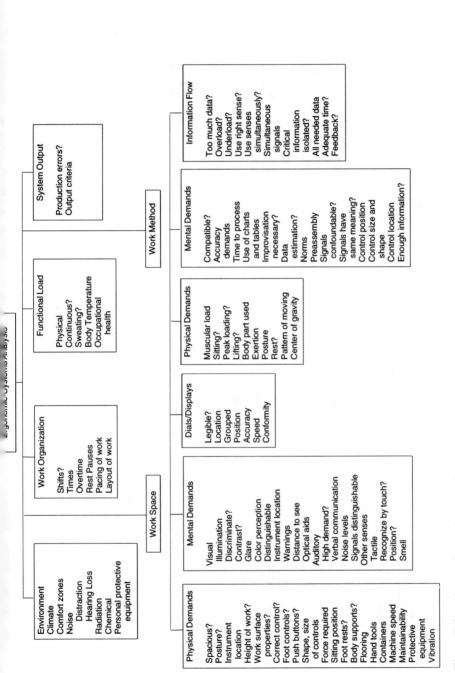

Figure 7-3. Human-factors (ergonomic-systems) analysis.

Table 7-1. Format for task allocation and analysis

Function (1)	Operate aircraft power plant and system controls
Task (2)	Control jet engine operation

Subtask (3)	Action Stimulus (4)	Required Action (5)	Feedback (6)	Task Classification (7)	Potential Errors (8)	Time (9) Allowable (9a)	Time (9) Necessary (9b)	Work Station (10)	Skill Level (11)
All Adjust engine r.p.m.	All Engine r.p.m. on tachometer	All Depress throttle control downward	6.1 Increase in indicated tachometer r.p.m.	All Operator task, aircraft commander	8.1 a. Misread tachometer b. Fail to adjust throttle to proper r.p.m.	10 sec.	7 sec.	aircraft commander's seat	11.1 Low

Source: Van Cott, H. and R. Kinkade, *Human Engineering Guide to Equipment Design.* Washington, DC: U.S. Government Printing Office, 1972.

2. Error *likely* to occur unless specific avoidance technique is utilized (probability greater than 0 but less than 0.5).
3. Error *unlikely* to occur even if specific avoidance technique is not utilized (probability of 0 or very close to it).

The system consequences of each equipment malfunction and human error should be described. An estimate of the seriousness of these consequences may be made on the following scale:

1. Hazard to personnel and/or equipment.
2. Degradation of system performance.
3. Degradation of subsystem performance.
4. Degradation of component performance.
5. Little effect on system performance.

The estimate of consequence seriousness or extent may be judged during the development phase by analyzing block and circuit diagrams relevant to the malfunctioning part of the data chain in which human error was made.

When these two analyses, the error analysis and the system consequences analysis, are considered together, the result is called an "error mode and effects" analysis. The designer must eliminate the possibility of an operator action that would have a catastrophic effect on personnel or equipment safety even if the event has a low probability of occurrence. On the other hand, the designer should not be concerned with eliminating those human errors or stylistic differences in performance that have little or no effect on other personnel, on equipment, or on system performance. There are several alternatives in eliminating or minimizing the probability of a human error occurring. The safest approach is to design the system so that the operator cannot commit the error. Alternate approaches that can be considered are: warning and caution labeling, procedure development, and training emphasis.

Van Cott and Kinkade also suggest the use of the questions that follow:[2]

1. Preliminary design phase
    a. Why is this system being sought?
       What mission will the system be expected to fulfill?
       More particularly, what is this new system expected to do that existing systems are not doing or are not doing well enough?
    b. How is the system to fulfill its mission?
       What are the stages of mission execution?
       What functions must be accomplished by the system at each stage?

---

2. Ibid.

    c. In what environments must the system function?
       What particular hazards will obtain?
       What stresses or demands are likely to be placed on the system?
    d. Who will benefit by system operation?
       Who will use the system?
       What kinds and numbers of operator and/or maintenance personnel are available?
    e. What are the major technological options?
       What alternative configurations are feasible?
       What particular resource or class of resource is most crucial to prospective system effectiveness?

2. Advanced design phase
    a. What functions should be assigned to human operator and support personnel?
       What conditions will impose peak task load on the operator or operators?
       What conditions (e.g., long periods of inactivity) will tend to degrade operator performance?
       What pattern of decision-action will occur at crucial mission stages?
    b. What information is required by operators (and/or support personnel) in order to fulfill their functions?
       What is the probable pattern of channels and of flow rate for this information?
       In what form (i.e., code, mode, format) will the information be most useful to the operator?
    c. How many humans are needed to man and support the system under normal and peak load condition?
       What special skills, capabilities or attributes are needed for effective operator performance?
       What special training, if any, will be required?
       Is such training feasible?
       What resources will be required to implement the training?
    d. How should the assigned functions be distributed among operator and support personnel?
       How should the work stations be arranged?
       What instrumentation is required at eack work station?
       How should this instrumentation be laid out?
    e. What specific devices, tools or controls are most appropriate to the pattern of task actions that will be imposed on operator and support personnel?
       What kinds of aids, guides, indicators, locks, interlocks, cover plates, etc. would be useful to facilitate correct actions and prevent operator errors?

Reducing System-Caused Error 105

What means are available to allow quick recovery or to maintain the safety and integrity of the system in the event of operator error or failure?
3. Mock-up to prototype fabrication phase
   a. What options are available for eliminating, combining, or amplifying any of the instrumentation?
   b. What will be the effect, if any, on human performance, safety, or morale of any proposed changes in configuration or instrumentation?
   c. What safeguards, if any, are required to insure adherence to the design plan and functional requirements of the system?
   What quality control procedures are required to insure the validity of human factors considerations in the final product?
4. Test and evaluation phase
   a. By what means can test and evaluation be made as realistic as possible in terms of the ultimate operator and support personnel and in terms of the operational conditions?
   b. What criteria of system and operator performance are logical in terms of the mission and functions assigned?
   What measurement procedures will yield data which are valid with respect to such criteria?
   What test instrumentation is required?
   c. What form of test design will yield unequivocal answers to questions of the effectiveness, operability and maintainability of the system?
   What is the most economical way of implementing the test design required?

Table 7-2 suggests which analysis techniques might be applicable for different levels of analysis.

There is one other analysis most appropriate today, the analysis necessary to ensure compliance with the governmental guidelines on cumulative trauma disorder (CTD).

CTD ANALYSIS

Cumulative trauma disorder analysis is required under OSHA's ergonomic guidelines. As indicated in Chapter 6, we do not really know how serious a problem cumulative trauma disorders (CTDs) are, as the figures vary widely. We are not even sure that CTDs are caused by work (repetitive motion).

Even so, the ergonomic guidelines require a CTD analysis on every job. The details of the CTD analysis are included in Appendix B. In effect, the CTD analysis

Table 7-2. Levels of analysis and some related techniques

Level of Analysis	Purpose	Applicable Technique
System	To determine effectiveness of system in performing a specified mission	Operations-research methods
Subsystem	To determine best way of meeting a specified requirement of the mission	Systems analysis integration matrix
Function	To determine best combination of components required to make up subsystem	Worker-machine system analysis Function analysis
Task	To determine best allocation of worker's capabilities to perform required functions	Task analysis Time-line analysis Logic models Information theory
Subtask	To determine best method of utilizing worker's capabilities to perform the assigned tasks	Operator-load analysis Operator-sequence diagrams Design theory Information-flow analysis
Element	To determine best method of utilizing worker's capabilities to perform assigned subtasks	Time-and-motion analysis Elemental-task analysis

Source: W. Woodson and D. Conover, *Human Engineering Guide for Equipment Designers*. Berkeley: University of California Press, 1965.

looks at each job, and what that requires in terms of strength, position, movement, repetitiveness, and so on, of the neck, shoulder, arms, wrists, fingers, trunk, back, and legs. Where predetermined standards are exceeded, the guidelines require action plans to meet the standards.

REFERENCES

Johnson, W. *The Management Oversight Risk Tree—MORT*. Washington, DC: U.S. Government Printing Office, 1973.

McFarland, R. Human factors engineering. *ASSE Journal,* February 1964.

Van Cott, H. and R. Kinkade. *Human Engineering Guide to Equipment Design*. Washington, DC: U.S. Government Printing Office, 1972.

Woodson, W. and D. Conover. *Human Engineering Guide for Equipment Designers*. Berkeley: University of California Press, 1965.

PART III | *Overload*

Chapter 8

Causes and Outcomes of Overload

As discussed in Chapter 3, overload comes from the improper matching of capacity with load, taking place in a given state. When this occurs a person may be overloaded or underloaded. Either is dangerous. The individual's state is important, as a person can be overloaded severely for a long period, and yet with the right attitude, or the right motivation, can survive and thrive. Without that attitude or motivation, only a small overload can be disastrous (the straw that breaks the camel's back).

Overload can be physical, physiological, or psychological. Because physical and physiological overload are dealt with in number of other places (OSHA standards, industrial hygiene texts and standards), we primarily look at psychological overload and its outcomes in this chapter, and at psychosocial factors on the job.

OVERTIME, SHIFT WORK, AND OTHER TRENDS

Overload is more prevalent today than ever before in organizations. Adrienne Burke describes it in this way:[1]

> Overtime. So much overtime it's making plants unsafe, some workers and shift work experts say.

---

1. Reprinted with permission from A. Burke, The push to produce, *Industrial Safety and Hygiene News*, December 1994.

> In several recent walkouts in the auto, steel and heavy equipment industries, labor leaders cited health and safety hazards in overtime work. Union representatives relate a plethora of health effects to overtime: repetitive strain injuries, severe stress, chemical sensitivities, and respiratory illnesses are a few. And fatigue invites catastrophes, not just around plant machinery, they say, but on the drive home too.
>
> Indeed, overtime hours are at a historic high now, according to the Bureau of Labor Statistics. In September, workers in manufacturing industries averaged 5.1 overtime hours weekly. A Bureau of National Affairs survey in July found that 90 percent of responding employers occasionally required overtime of workers. Of continuous operations (running 24 hours a day, 7 days a week), nearly half admitted to frequently mandating extra hours.
>
> Stories pointing out overtime dangers abound. Interviews for this report uncovered these: A maintenance worker is so fatigued he fumbles around looking for a wrench and finds it in his other hand. After four days of 18-hour shifts, a utility line worker electrocutes himself. A nurse accidentally sticks herself with an HIV-infected needle in the twelfth hour of a 12-hour shift.
>
> Of more than a dozen safety managers interviewed for this report, most are not convinced that overtime triggers more accidents. A definitive study has yet to be done. And, until now, no one seems to have cared.
>
> But circumstances today are unique: Trimmed-down companies are determined to stay lean; employers still reeling from heavy staff cuts hesitate to hire again so soon. Benefits costs and training requirements now make paying time-and-a-half a much better deal than hiring a new employee. So overtime is being mandated like never before.
>
> To be sure, few who protest excessive overtime suggest that a hiring boom will prevent accidents. Labor leaders and safety professionals alike admit that inexperienced workers have more accidents than veteran workers.
>
> What's more, plenty of workers like overtime. Many rely on the additional cash flow. Some workers who support families even say they feel guilty denying extra work if it's offered.
>
> Trouble is, more and more, overtime isn't offered. It's required.

Scott Robertson, writing in *New Steel*, traces current (1994–95) trends in the steel industry with respect to serious injuries:[2]

> The steel industry is booming again. But with increased production has come increased numbers of deaths as workers sometimes struggle to keep up with higher demands.
>
> Twenty-one members of the United Steelworkers of America (USWA) died in the first half of 1994, the USWA reports. Three supervisors or contractors in USWA-represented plants also died in that time.

---

2. Reprinted from S. Robertson, Increased production, increased fatalities, *New Steel*, October 1994.

If that pace continues, the USWA will have its highest fatality rate on record, an average of about 20 fatalities for every 100,000 workers. Seven USWA members died during all of 1993.

The union is investigating four trends to see their impact on the number of fatalities and injuries in mills.

First is the high capability-utilization rate . . . "a bottleneck of people taking extraordinary, and sometimes unsafe, steps to maintain that capacity."

Second is an increase in overtime at plants. In the early decades of this century, it was not uncommon for maintenance employees to work steady double shifts that added up to 84-hour work weeks, Wright says. Some maintenance people might be approaching those levels again.

"Overtime is different than it was 10 years ago," Wright says. "In the early 1900s, on the second part of a shift, the maintenance guy might not work at all. They were on call, but they could rest. Now the companies keep up better maintenance schedules and the maintenance workers are doing more double shifts. We think the impact of the overtime is much greater than the hours suggest."

Third is the number of job changes. Many of the deaths, the union believes, result when longtime employees begin unfamiliar tasks, Wright says. "Traditionally, the way workers learned their jobs was to go into the plant and talk to Joe, and have him show you the ropes," Wright says. "That situation is changing, but not fast enough."

Fourth is the influx of new technology in the mills. New processes create new ways of doing jobs. As workers take time to learn the new ways, they could be more vulnerable to accidents.

Experts on sleep and shift work say studies do link accidents to work hours. However, it's not the number of hours, but when they're worked that matters, they say. Harvard sleep expert, Dr. Martin Moore-Ede, points to examples in the railway industry: "If you examine all the major accidents over the last few years, there's little correlation with the number of hours the crews worked. But there's a strong correlation with the time of day accidents occurred."

The dangerous time, according to Moore-Ede and other experts, is between 4 a.m. and 6 a.m. Cognitive error is most common then. In fact, Moore-Ede says, most people can work a shift from 7 a.m. to 3 p.m. and then stay for another shift without too much trouble. "But if you're working the evening shift and are held over for the night to work during the time you would normally be sleeping, it could have very dramatic effects."

A lot of research has been done on shift work and sleep. Sandy Graham, in *Safety & Health*, writes:[3]

---

3. Reprinted with permission from S. Graham, Are your employees dead tired? *Safety & Health,* March 1995.

Many of the 22 million shift workers in the United States, particularly those who work rotating shifts of days, then afternoons and then nights, suffer from sleep disorders. The National Commission on Sleep Disorders Research reports that insomnia, the inability to sleep or sleep well, affects an estimated 40 percent to 80 percent of shift workers. On average, a shift worker sleeps two to four hours less a night than a day worker in the same age group. Shift work is like going on a nonstop, round-the-world trip from one time zone to the next.

Sleepy workers aren't productive workers either. The commission concluded that sleep problems cost the United States about $15.9 billion a year in direct costs for medical care or self-treatment. Companies may lose as much as another $150 billion from indirect costs of compromised productivity.

The human body is programmed for regular sleep/wake cycles. Its internal clock—the hypothalamus of the brain—calls for a long period of nighttime sleep. According to the National Commission on Sleep Disorders, we're generally sleepiest between 2 a.m. and 5 a.m.

There's also a dip in alertness between 2 p.m. and 3 p.m. Many cultures allow for that decreased alertness with some form of rest break, such as a siesta. Western cultures tend to ignore the body's call for afternoon rest; in the United States, "napping means lazy, unmotivated" (S. Graham).

REACTIONS TO OVERLOAD

The concept of overload is nothing new. Industrial hygiene Maximum Allowable Concentrations (MACs) and Threshold Limit Values (TLVs) are nothing more than a definition of "how much is too much." And these quantities are typically well defined. It is in the psychological area that overload seems fuzzy to most of us in safety.

Perhaps the best way to look at overload is to define it as when there is just "too much"; for instance, where there is too much tension, too much conflict, too much nonsense to deal with. A certain amount of all this is okay and not a problem. In fact, there is some real value in building some tension into the organization:

- Tension may speed executive development: the learning, the understanding, and the internalizing.
- It can be a stimulus to imagination and performance.
- It is an integral part of competition, so important to American business.

Conflict

The question is when is there too much tension, and when does it become unhealthy to the organization? Chris Argyris has discussed the innate conflict between mature human needs and what the organization can do to meet those needs. Argyris's conflict theory suggests rather strongly that the more classical management is in style, the more conflict there will be.

The innate conflict Argyris discussed is not the only conflict people in organizations feel. There are normally several more types. Typically there will be a felt conflict by a subordinate when the boss uses a leadership style inappropriate to that subordinate's needs. There may be a felt conflict when the subordinate's input is not asked for, is not listened to, or is ignored. In short, conflict will be perceived in many situations where human needs are not being met. And the problem probably will continue to worsen. As workers become even better educated, get a taste of participation and want more, and experience recognition from one boss and come to expect the same in other job settings and do not get it, more conflict will be felt.

Hostility

Frederick Herzberg is another one of the many behavioral theorists to discuss this problem. He uses the word "hostility" to describe it and suggests that we must learn to manage the hostility. Hostility is the result of biochemical body changes, the secretion of neuro-hormones at nerve fiber connections. This surplus must be neutralized, so we take things like Rolaids. At the behavioral level we transform the hostility into something we can handle by walking it off or punching something. When we cannot do this, we internalize the feelings, and this leads to trouble.

Most researchers have found that nothing in industry is as damaging to health, performance, and the company's economic good as stress, apprehension, fear, and hostility.

Herzberg describes the physiological responses to hostility and how they end up in psychological response: displacement or internalization. Perhaps an easy way to picture this is to think in terms of the reactions of workers to management's actions. In physics we learned a simple, but important, law: for every action, there is an opposite reaction. While the law is stated for our physical universe, it has broad implications in our psychological universe.

When there is an action against us psychologically, we react in some way to that action. The action against the worker is the process of changing attitudes, perhaps even values, to be in tune with the "acceptable" behaviors, attitudes, and values of the organization. The process is hard on most people. It creates stress, which results in strain on our physical condition. It forces us into mediocrity; it destroys creativity. If we live through it (that is, do not quit) "they" succeed in socializing us until we "fit" and become promotable.

Flight or Fight Symptoms

But the law of reaction above suggests that such action against people does not usually go unanswered. People will react in some fashion. Typically the reaction will come in one of two ways. They will react against the action (fight back), or they will attempt to escape the action (the fight or flight response).

Some options will be healthy to the organization, some unhealthy. Figure 8-1 describes some of the options. There are no doubt more that the reader can add to each list. In looking at the lists in Figure 8-1, a few things seem obvious:

- There are a lot more unhealthy reactions than healthy ones.
- Many of those that are healthy to the worker are unhealthy to the organization.
- It is healthy for the worker to fight (at least in this scheme).
- Workers seem to have a wealth of options to choose from.
- Quite a few of their options are related to the organization's safety results.

Figure 8-1 suggests that the simplest place to start is with the column "healthy flight," as there is only one reaction there: *get promoted*. This is normally healthy for both the individual and the organization. In the list of the symptoms of conflict this is the only positive one (from management's viewpoint): get promoted and work up in the organization to become a manager. Why is promotion perceived as "flight"? Because it is the process of avoiding conflict by joining the other side. If this is the only healthy reaction to both sides, why is it not used more often? Probably for two reasons:

1. A lot of hourly workers have experienced enough stress and feel that they need no more; and they have seen what the first level of management consists of: overwork, hassles, shifting management goals, more work than time to do it. In terms of overload, in most organizations the most overloaded person clearly is that first line supervisor, who almost always has 80 hours of work for every 40 available hours—and everybody knows it. Promotion only adds to the overload problem.
2. In many organizations a caste system has developed; it is difficult to come up through the ranks. College graduates are hired as supervisors as the first step in their career path. This is the old military system of officers and enlisted men. A number of industries are structured this way. Where this is true, the whole option of a "healthy flight" is closed off to the worker, who must either fight or choose an unhealthy reaction.

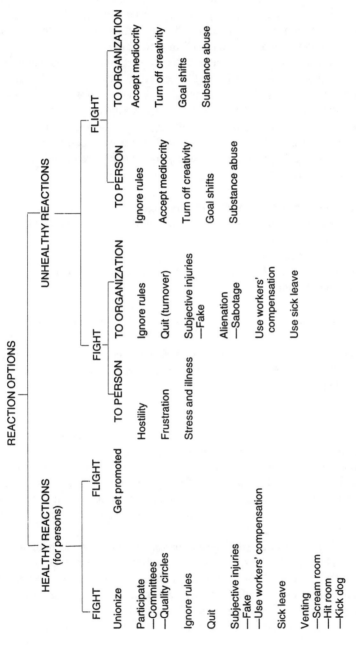

Figure 8-1. Fight or flight symptoms.

The Healthy Fight Reactions

Here we have a number of options, all healthy for the individual worker but most unhealthy to the organization. To *unionize* is one option—to formally band together to fight, or more accurately to formally choose to be represented in the fight by a different organization, the union. Does this solve the conflict? Yes and no or perhaps, "depending upon the situation." In some cases it heightens the conflict felt by the worker; it stiffens the battle lines. It needs to be said, however, that a strong and militant union is usually well deserved by past management actions.

A particularly healthy fight happens when workers choose to *participate* to make a situation better through whatever mechanisms are available to them: safety committees, quality circles, ad hoc committees, or a management that simply asks for help. This is perhaps the best of all worlds because everybody wins.

Every behavioral theory and every managerial theory says participation ought to work, and it does. It works unless:

- Management dictates how and when people can participate.
- Management uses participation as a new and fancy way of getting the workers to "jump through new hoops."
- Management asks for, and then ignores, input from hourly workers.
- Management actions in the past have been such that workers "know" they really do not want help.

Some healthy reactions for the worker that are unhealthy for management are:

- *Choosing to ignore rules* of management because of a perception (often learned) that "they don't know" or "they don't care," or "they haven't earned the right to say that to us." This is a fight reaction, as it is a distinct (conscious or unconscious) effort to say, "We don't care what you say; we'll do it our way."
- *Choosing to quit.* This is often healthy for the worker, as he or she has a chance to improve the situation in a new job.
- *Choosing* (consciously or unconsciously) *to use the system* through a "pseudo-injury," a fake injury, a "Monday morning" injury, or using sick leave to get time off. This is healthy for the worker as it fulfills the need for time off, to escape temporarily the pain felt on the job. It is clearly unhealthy for the organinzation.
- *Venting the frustration* (letting off steam). As Herzberg was quoted earlier, this can be unhealthy to others: spouses and dogs that get kicked a lot. But it is healthy to the worker—displacing is better than internalizing. Some organizations are dealing with this today by providing opportunities for venting. Some organizations have provided a "scream closet" where an employee can go to vent the rage, the hostility, and the conflict. Some Japanese companies

offer a room with punching bags, each bag having the face of a supervisor or manager (might be a good tool for performance appraisals of managers—the most beat-up achieves the lowest rating). More common approaches are counseling assistance through Employee Assistance Programs (EAPs).

The Unhealthy Reactions

We have a long list of unhealthy reactions, some of which have already been mentioned. The first might be the worst. The fight reaction of hostility and frustration with no outlet leads to stress-induced illness. Unhealthy flight reactions are equally serious. For instance:

- *The forcing of mediocrity.* As a person is socialized through time and the organized efforts of management, both abilities and the will to work change. Mediocrity is defined as less than top performance. When workers produce less than they can, or managers accept less from themselves or their subordinates than they can get, these are expressions of mediocrity.
- *The destruction of creativity.* Just as mediocrity (less than top performance) becomes acceptable, so also does acceptance of doing things the way they have always been done. Instead of using our innovative thoughts, our creativity to improve the situation, we accept "the way it is."
- *Goal shifts.* Workers experience goals shifts—from buying into the organizational goals to buying out. Why? Because over time they learn that meetings are boring, that the approach is negative, and that management does not want their input. To maintain their psychological health, they "buy out," but in the long term this is unhealthy to them as well as to the organization.
- *Alienation, strikes, and sabotage.* Perhaps the most unhealthy reaction to the organization is alienation so severe that it leads to strikes and even sabotage. While uncommon, this does exist as a severe reaction to perceived management abuses.
- *Substance abuse.* A similar severe unhealthy reaction ends up in drug or alcohol abuse.

PSYCHOSOCIAL FACTORS ON THE JOB

A major study of organizational signs of stress was made by the ILO/WHO Committee on Occupational Health in 1986. It studied assessment and consequences of psychosocial factors at work.

Factors studied were the interactions between and among work environment, job content, organizational conditions and worker capacities, needs, culture, and other considerations that may influence health, work performance, and job satisfaction (see Figure 8-2).

The results included the following:

- With respect to the physical work environment:
  - Workers reported variation and chemical exposures as the most harmful perceived stressors.
  - Work overload was often cited as bringing on health complaints and psychological problems.
  - Underload and lack of control were cited most often in certain jobs (nuclear power plants notably).
  - Shift work was cited as a major stress producer.
- With respect to worker's role, these were cited most:
  - Role ambiguity.

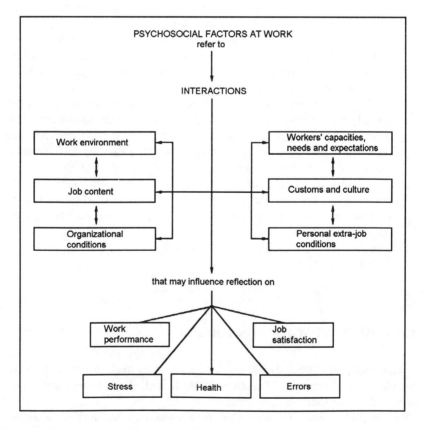

Figure 8-2. Psychosocial factors at work.

- Role conflict.
- Conflict with organizational boundaries.
- Responsibility for others.
- Regarding worker's participation, those cited were:
  - Climate, including politics, and lack of participation in decision making.
  - Unjustified restrictions on behavior.
- Determination of relationships at work, particularly with superiors.
- Regarding implementation of changes, mentioned most were:
  - Lack of communication before and during the change.
  - New methods not assessed in human terms.
  - Not enough supportive measures provided during change.
- Regarding industrialization, mentioned most was:
  - Lack of training.
- On new technology the problems were:
  - Repetitiveness and boredom.
  - Lack of contact with the customer.
  - Monotony.
  - Computer breakdowns.

The psychosocial factors on the job determine the amount of stress workers experience, which influences injuries, illnesses, and cumulative trauma disorders, which affect worker's compensation claims. John Kamp describes the process:[4]

> Employees engaged in physical work have always suffered aches and pains. However, until 10 years ago, these employees could take only two courses of action: 1) keep working despite the pain or 2) find another occupation. Given the "self-diagnosed" nature of disability in most cumulative trauma cases, however, these employees now have a third choice: filing a WC claim.
>
> Psychological factors, job attitudes and personal traits play a key role in whether this becomes the option of choice. The role of job attitudes is straightforward. Consider Susan and Dorothy, two employees who perform the same physical work in neighboring factories. In addition to equal physical exposure, these employees also suffer an identical, moderate level of pain.
>
> Satisfied Susan loves work—she gets along well with her supervisor, her best friends are at work, and she feels her employer cares about her welfare. Conversely, Disenchanted Dorothy hates work—her boss always berates her, her work unit is full of animosity and tension, and she feels her em-

---

4. Reprinted with permission from J. Kamp, Worker psychology: safety management's next frontier, *Professional Safety*, May 1994.

ployer is merely using her. Clearly, continuing to work despite the pain is a more attractive option to Susan, while staying home and collecting workers' compensation has greater appeal to Dorothy.

Certain personal traits increase the likelihood that an employee will file a cumulative trauma claim, even when another employee with a similar pain level and job attitude would not. Hypochondria, the tendency to constantly adopt a "sick person" role, is one such trait. This behavior is reinforced by attention from doctors, family, friends, etc. Being "out" on workers' compensation provides an opportunity to be lavished with attention.

Another trait that contributes to cumulative trauma claims is a poor work ethic. Certain individuals view work as a temporary annoyance (rather than one of life's responsibilities) to be tolerated only until a more desirable source of funding becomes available. When these individuals experience work-related pain, workers' compensation offers an attractive way to "make a living." (In the extreme, the trait of dishonesty leads some to file fraudulent claims in the absence of pain or injury.)

RESEARCH EVIDENCE

Research supports the role of psychological factors such as stress and job attitudes in WC injuries. For example, a prospective study of 3,000 Boeing workers found that, apart from a history of back problems, the strongest predictors of reported back injuries during a four-year follow-up period were not physical factors, but rather worker perceptions and psychological traits. "Subjects who stated that they 'hardly ever' enjoyed job tasks were 2.5 times more likely to report a back injury ($p = .0001$) than subjects who 'almost always' enjoyed their job tasks" (Bigos et al. 1).

A NIOSH study examined factors related to upper extremity musculoskeletal disorders and symptoms among U.S. West Communications employees who used video display terminals (VDTs). Psychosocial variables associated with disorders and symptoms included increasing work pressure, surges in workload, lack of job diversity, limited decision-making opportunities, uncertainty about job future, fear of being replaced by computers, and poor supervisor and co-worker support. According to the researchers, "This study adds to evidence that the psychosocial work environment is related to the occurrence of work-related upper extremity musculoskeletal disorders" (Hales et al. 4).

St. Paul Fire and Marine Insurance Co. has conducted studies of relationships between stress, job attitudes and WC claims. These studies have utilized Human Factors Inventory (HFI), an employee survey that measures various "human risk factors" in an organization's workforce. Factors include work stressors (i.e., excessive workload), personal life stressors (family problems) and lifestyle risks (substance abuse) that increase

Causes and Outcomes of Overload 121

	Per 100 employees, past year		Average HFI Scale Scores					
Location	WC Freq.	WC Claim Costs	WL	SP	WR	PP	PLS	JS
Low claims	1.9	$921	59	43	41	52	44	51
High claims	15.5	$192,770	77	58	59	66	55	65

Note: HFI scores of 50 are average; higher scores are more unfavorable.
WL = Workload; SP = Supervision; WR = Working Relations;
PP = Personnel Practices; PLS = Personal Life Stressors;
JS = Job Satisfaction

Figure 8-3. HFI scores and WC data. (Reprinted with permission from J. Kamp, Worker psychology, *Professional Safety,* May 1994.)

organizational costs through accidents, injuries, health problems, and lowered productivity, quality, and customer service.

Figure 8-3 shows HFI results and WC data for low-claims and high-claims locations of a nationwide copy machine distributor. As shown, the 15 low-claims locations reported minimal WC frequency rates and costs in the year prior to surveying employees. The three high-claims locations reported an average standard frequency rate eight times higher than the low-claims locations, with average costs nearing $200,000 per 100 employees.

High-claims locations scored significantly worse ($p < .05$) on many HFI scales, indicating greater workload pressures, poorer supervision and working relations, more dissatisfaction with personnel practices, greater personal life stress and greater overall job dissatisfaction. Within this company, therefore, high levels of work, personal stress and unfavorable job attitudes were associated with higher WC claim rates and costs.

Figure 8-4 shows HFI results and WC data for four California locations of a nationwide cable television company. California is the nation's leader in compensating purely mental job injuries (so-called stress claims). Therefore, by separating these locations' WC claims rates according to stress and non-stress claims, one can uncover a clear picture of trends. Rates reflect claims spanning the 15 months preceding the survey.

Not surprisingly, two locations reporting high levels of work stress (C and D) experienced higher stress claim rates. However, these locations also reported appreciably higher rates for nonstress claims. These findings support the argument that excessive work stress contributes to both physical and mental injury claims.

Scientific research verifies the "real-world" experience that psychological factors play a key-and expanding-role in WC claims. Although ac-

WC Claim Freq., past 15 mo.

Location	Stress	Non-stress	Total	Avg. HFI[a]
A	2.5	9.4	11.9	48
B	1.6	17.9	19.5	51
C	8.2	28.3	36.5	61
D	12.6	28.8	41.4	62

[a]Average score across the 6 HFI Work Stressors scales.
Note: HFI scores of 50 are average; higher scores are more unfavorable.

Figure 8-4. HFI scores and WC data. (Reprinted with permission from J. Kamp, Worker psychology, *Professional Safety,* May 1994.)

knowledging this fact is a positive step, the greater challenge lies in taking action.

When present, the four psychological factors described here—work stress and personal life stress (traumatic accident model), and unfavorable job attitudes and personal traits (cumulative trauma model)—increase the likelihood that an individual will be involved in a WC claim. Can an organization realistically expect its workplace to be completely free of stress, job dissatisfaction or "difficult" personalities? Probably not. However, an organization *can* better manage these factors and, thus, improve its overall safety performance.

REFERENCES

Argyris, C. *Personality and Organization.* New York: Harper, 1957.
Bigos, Stanley J. et al. A prospective study of work perceptions and psychosocial factors affecting the report of back injury. *Spine* 16:1–6, 1991.
Burke, A. The push to produce. *Industrial Safety and Hygiene News,* December 1994.
Ewing, D. Tension can be an asset. *Harvard Business Review,* September–October 1964.
Gellerman, S. In *Safety Management—A Human Approach,* 2nd ed., by D. Petersen. Goshen, NY: Aloray, 1988.
Graham, S. Are your employees dead tired? *Safety & Health,* March 1995.
Hales, Thomas et al. *Health Hazard Evaluation Report: HETA 89-299-2230, U.S.*

West Communications. Cincinnati, OH: National Institute for Occupational Safety and Health, 1992.

Herzberg, F. In *Safety Management—A Human Approach,* 2nd ed., by D. Petersen. Goshen, NY: Aloray, 1988.

ILO/WHO Committee on Occupational Health. *Psychological Factors at Work.* Geneva; ILO, 1986.

Kamp, J. Worker psychology: safety management's next frontier. *Professional Safety,* May 1994.

Kohler, Stacey and John Kamp. American Workers Under Pressure, Technical Report. St. Paul, MN: St. Paul Fire and Marine Insurance Co., 1992.

Likert, R. *The Human Organization.* New York: McGraw-Hill, 1967.

NIOSH Director Millar speaks out on trends in the health and safety industry. *Occupational Health & Safety,* pp. 26–27, October 1992.

Robertson, S. Increased production, increased fatalities. *New Steel,* October 1994.

Yariger, A. Stress—the father of disaster. *Industry Week,* September 20, 1971.

Chapter 9

Capacity

In considering the capacity aspect of overload, we must examine quite an array of factors. Probably the best starting point is where we left off in Chapter 6 with our discussion of the worker–machine system. McFarland's diagram, Figure 6-2, shows the human subsystem as consisting of sensory organs, a central nervous system, and effectors. In Wood's diagram, Figure 6-3, the human subsystem is shown as somewhat more complex, consisting of a mind with (1) an environmental input analyzer, (2) feelings and thoughts, (3) an analysis and decision-making capability, and (4) a behavior identifier. In addition to the mind, Woods shows internal inputs, control, and output as parts of the human subsystem. In effect, both McFarland and Wood are dealing with what might be considered three subsystems of the human being—a sensing subsystem, a responding subsystem, and an information-processing subsystem.

HUMAN SUBSYSTEMS

Information Processing

The three human subsystems are shown in some detail in Figure 9-1. Van Cott and Kinkade describe the various components of the sensing, responding, and information-processing subsystems and also identify a fourth one, which they call the storage subsystem. They suggest that the information-processing subsystem consists of four components: (1) pattern recognition, (2) decision making, (3) adaptive processes, and (4) timing and time sharing. This information-pro-

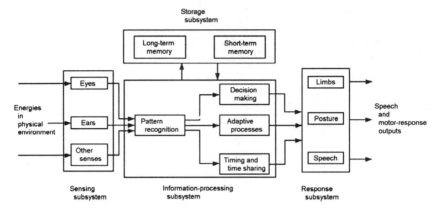

Figure 9-1. The human information-processing system (From H. Van Cott and R. Kinkade, *Human Engineering Guide to Equipment Design.* Washington, DC: U.S. Government Printing Office, 1972.)

cessing subsystem will be discussed in more detail in the next chapter, as it is often a major factor in load.

Sensing

Much has been written about the sensing subsystem. Sinaiko has listed the various sensing mechanisms of human beings and the physical energies that stimulate them (Table 9.1), Deatherage has compared the relative merits of auditory and visual presentations (Table 9.2), and McCormick has experimented to determine the number of absolute judgments that can be made by a person of specific stimuli (Table 9.3). A great deal of research has been done on the senses, their functions and limits, and the relation of these to accidents.

Vision

The subject of vision includes blindness, adaptation, color blindness, depth perception, vision, and illusions.

The eye, the most important human sensor, has been neglected in accident research. There is some indication that there is a relation between poor visual function and accidents, but better studies are needed.

Only two industrial studies specifically have investigated the relation between accidents and visual performance of workers. One was a study of 1,384 miners with 1,265 accidents over a period of two and one-half years. No strong relation between poor visual acuity and accident repeaters was found. This might be expected when all types of accidents are investigated.

Capacity

Table 9-1. Human senses and the physical energies that stimulate them

Sensation	Sense Organ	Stimulation	Origin of Stimulation
Sight	Eye	Some electromagnetic waves	External
		Mechanical pressure	External or internal
Hearing	Ear	Some amplitude and frequency variations of the pressure of surrounding media	External
Rotation	Semicircular canals	Change of fluid pressures in inner ear	Internal
	Muscle stretching	Muscle stretching	Internal
Falling and rectilinear movement	Semicircular canals	Position changes of small, bony bodies in the inner ear	Internal
Taste	Specialized cells in tongue and mouth	Chemical substances dissolvable in saliva	External on contact
Smell	Specialized cells in mucous membrane at top of nasal cavity	Vaporized chemical substances	External
Touch	Chiefly skin	Surface deformation	On contact
Vibration	Several organs	Amplitude and frequency variations of mechanical pressure	On contact
Pressure	Skin and underlying tissue	Deformation	On contact
Temperature	Skin and underlying tissue	Temperature changes of surrounding media or of objects contacted	External and on contact
Cutaneous pain	Unknown, but thought to be free nerve endings	Intense pressure, heat, cold, shock, chemicals	External on contact
Subcutaneous pain	Thought to be free nerve endings	Extreme pressure and heat	External and on contact

Source: Adapted from H. Sinaiko, Selected papers on human factors in the design and use of control systems. Chicago: National Safety Council, 1970 Reprint.

In another study, researchers attempted to select accidents that could possibly have been caused by low visual performance, and found that the accident-free did in fact have superior vision.

Hearing

Much has been written on hearing loss caused by noise, but few studies have been made to see if hearing defects cause industrial accidents. One study showed that the deaf had slightly fewer accidents, and another reported that 60 deaf

Table 9-2. Relative merits of auditory and visual presentations

Use Auditory Presentation if:	Use Visual Presentation if:
Message is simple	Message is complex
Message is short	Message is long
Message will not be referred to later	Message will be referred to later
Message deals with events in time	Message deals with location in space
Message calls for immediate action	Message does not call for immediate action
Receiving location is too bright	Receiving location is too noisy
Person's job requires him or her to move continually	Person's job allows him or her to remain in one position
Visual system of person is overburdened	Auditory system of person is overburdened

Source: Adapted from B. Deatherage, Auditory and other sensory forms of information presentation, in H. Van Cott and R. Kinkade, *Human Engineering Guide to Equipment Design,* Washington, DC: U.S. Government Printing Office, 1972.

Table 9-3. Number of absolute judgments that can be made by a human being of specific stimuli

Sensory Modality and Stimulus Dimension	Number of Levels that Can Be Discriminated on Absolute Basis
Vision: single dimensions	
Pointer position on linear scale	9
Pointer position on linear scale	
Short exposure	10
Long exposure	15
Visual size	7
Hue	9
Brightness	5
Vision: combinations of dimensions	
Size, brightness, and hue*	17
Hue and saturations	11–15
Position of dot in a square	24
Audition: single dimensions	
Pure tones	5
Loudness	5
Audition: combination of dimensions	
Combination of six variables**	150
Odor: single dimension	4
Odor: combination of dimensions	
Kind, intensity, and number	16
Taste	
Saltiness	4
Sweetness	3

Source: From *Human Factors in Engineering and Design* by E. McCormick. Copyright © 1976. Used with permission of McGraw-Hill Book Company.
*Size, brightness, and hue were varied concomitantly, rather than being combined in the various posssible combinations.
**The combinations of six auditory variables included frequency, intensity, rate of interruption, on-time fraction, total duration, and spatial location.

persons employed in a variety of jobs had no "lost-time" accidents over a period of three years.[1]

Motion Sensors

While there is no research about motion sensors, we do know that in most industrial situations unusual problems with balance organs will seldom arise because the person is usually standing on the ground; but if any rapid or unusual acceleration of vehicles or conveyers or self-propelled head movements should occur, close or precise work should not be required for a few seconds afterward.

Stability sensors are intimately related to the whole process of self-propelled body movement. Thus disorders and misinterpretation can lead to falls.

Skin Sensors

Under the surface of the skin there are special sensors for touch, temperature, pain, and pressure. There has been no research to correlate these sensors with accidents.

Taste and Smell

Taste and smell are considered to be unimportant with respect to accidents, except to those working with toxic chemicals. There have been no studies in this area. Generally, very little research has been attempted to uncover sensory errors as causes of accidents.

Responding

Considerable research has been done on the response subsystem.

Alertness

It is hypothesized that accidents are more likely to occur when a person is experiencing high fear arousal because he or she is unable to coordinate all the necessary information and act correctly on it. If the fear arousal is low, accidents can occur because the person is not observing the environmental clues and thus

---

[1] B. Deatherage, Auditory and other sensory forms of information presentation, in H. Van Cott and B. Kinkade, *Human Engineering Guide to Equipment Design*. Washington, DC: U.S. Government Printing Office, 1972.

cannot act on them. Optimum alertness is thus vital for efficiency of output and for safety.

Risk Taking

Tom Rockwell attempted to study risk-taking behavior in a controlled laboratory setting. He found that a person's judgment of risk is not directly related to the hazard as measured by his or her performance capabilities; skilled workers take fewer risks than unskilled; younger persons take more risks; females take considerably fewer risks than males. These laboratory tests did not reveal any correlation between risk taking and various psychological or biographical measures, although Rockwell did discover that the high-risk taker tends to be a person with a high anxiety level, high sociability, and low emotional stability.

Intelligence

Several researchers have found that accident frequency is unrelated to intelligence as long as the worker's intelligence is adequate for the situation.

Sensory Motor Tests

Measures of muscular coordination have been shown to be related to certain types of accident situations. It seems reasonable to suppose that clumsiness, inadequate skill, slowness of response, and defective sense organs contribute to accidents. Individuals low in motor ability are unable to get out of accident situations without sustaining an injury; also, in many cases, they do not have the skill or sensory acuity to avoid certain kinds of accident situations. Here are some research findings.

Reaction Time. Reaction time seems to be a more important consideration in some types of work than in others. Consequently, it is necessary to analyze the requirements of certain jobs in order to determine the importance of reaction time to safe and efficient performance in those jobs.

According to John Larson, tests can show whether a worker falls above or below the minimum safety requirements. But reaction time seems to involve other factors, such as manual or manipulative dexterity, rhythm (or timing), and visual acuity. Quick reaction alone will not necessarily prevent injury; the worker must also react in the right way and at the proper moment. To do these things, of course, the worker must foresee an imminently dangerous situation. Where it is purely a matter of vision, reaction time and dexterity seem to be the most important qualities. But when anticipation rather than visual acuity alone is involved, fast reaction time appears to be less significant. The importance of reaction time per se in accident

Capacity   *131*

causation depends largely on the nature of the work involved. Moreover, anticipation and, therefore, attention seem to reduce the need for a worker to have quick reaction time.

Visual Acuity. In one study of visual acuity, 828 workers in 12 job categories were studied and divided into two groups: those whose visual skills met the requirements of the visual pattern for the job, and those whose visual skills did not. From the accident data, workers in each job category were split into those who had two or fewer accidents in six to eight months, and those who had three or more accidents in the same period.

After comparing the two groups, the researchers concluded that vision is one of the factors related to accidents, but that many other factors are involved. The relationship shown in the study was not sufficiently high to account for all industrial accidents, but it showed that, among workers whose visual skills are adequate for the tasks they are performing, there is a higher percentage of workers whose accident frequency is low. Again, a worker's level of visual acuity by itself is not enough to predict accidents.

In another study, 40 operators were evaluated who represented a cross section of a department of 79 female workers. After administering three motor manipulation tests and two visual perception tests, the researchers compared 23 accident-prone workers with 17 accident-free workers. No significant relationship existed between an accident index and the several sets of test scores, but the accident-index figures tended to be high when the scores on the three motor tests were higher than the scores on the two perception tests, and vice versa. The conclusion was that individuals whose levels of muscular reaction are above their levels of perception are prone to more frequent and more severe accidents than those individuals whose muscular-reaction levels are below their perceptual levels. This finding was qualified by the suggestion that accident proneness probably arises from several factors and not from a single factor.

TOTAL HUMAN SUBSYSTEM CONSIDERATIONS

Besides considering the individual subsystems of sensing, information processing, and responding, we must also consider the human subsystem as a whole, for there are a number of variables that will and can influence human beings' total performance by reducing their capacity.

One reason for human error is the fact that we often use people to perform functions that could be better performed by a machine. Woodson and Conover suggest where each excels (Table 9-4). Beyond that, there is a whole area suggested in our model of natural endowment. Endowment refers to the abilities and strengths

Table 9-4. Human workers vs. machines

Human Workers Excel in	Machines Excel in
Detecting certain stimuli of low energy levels	Monitoring (both people and other machines)
Sensing an extremely wide variety of stimuli	Performing routine, repetitive, or very precise operations
Perceiving patterns and making generalizations about them	Responding very quickly to control signals
Detecting signals at high noise levels	Exerting great force, smoothly and with precision
Storing large amounts of information for long periods, and recalling relevant facts at appropriate moments	Storing and recalling large amounts of information in short time periods
Exercising judgment when events cannot be completely defined	Performing complex and rapid computation with high accuracy
Selecting own inputs	Being sensitive to stimuli beyond the range of human perception (such as infrared, radio waves)
Improvising and adopting flexible procedures	
Reacting to unexpected low-probability events	Doing many different things at one time
Applying originality in solving problems: that is, coming up with alternate solutions	Reasoning deductively, from generalities to specifics
Profiting from experience and altering course of action	Being insensitive to extraneous factors
Performing fine manipulation, especially where misalignment appears unexpectedly	Operating very rapidly, continuously, and precisely the same way over a long period
Continuing to perform even when overloaded	Operating in environments that are hostile to human beings or beyond human tolerance
Reasoning inductively, from specifics to generalities	

Source: Adapted from W. Woodson and D. Conover, *Human Engineering Guide for Equipment Designers.* Berkeley: University of California Press, 1966.

a workers starts with. There is a whole range of human differences. Some workers are built more strongly and can physically handle more difficult work; some workers are physiologically stronger and can stand more environmental stress; some workers are psychologically stronger and can handle more psychological stress. Some aspects of natural endowment can be identified during the selection by medical examination. Physiological and psychological endowments are difficult and perhaps impossible to assess before selection and placement.

In *Job Demands and Worker Health,* a NIOSH (National Institute for Occupational Safety and Health) research report conducted by the Institute for Social Research at the University of Michigan, the model in Figure 9-2 was presented. It identifies the relationships among the environment (both actual and subjective), the person–environment fit, the person and his or her social support, and the person's responses to a situation. The report also presents and discusses the model shown in Figure 9-3, which pertains to the causes of occupational health.

Capacity 133

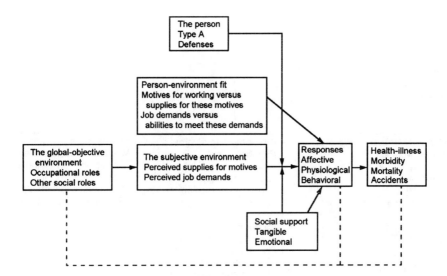

Figure 9-2. A theory about the effects of social stress on health. (Adapted from NIOSH, *Job Demands and Worker Health,* HEW Publication No. 75-160. Washington, DC: U.S. Government Printing Office, 1975.)

In the model diagrammed in Figure 9-3, the traits of the person are considered in relation to his or her subjective environment (job stress), and with actual versus desired levels of work load, responsibility, complexity, and ambiguity. These stresses together create certain kinds of responses; psychological, physiological, and behavioral, which eventually result in either illness or good health.

Psychological and Personality Traits

Intelligence

Research shows that intelligence bears little relation to accidents, except where accidents are due to failures in judgment. In these cases, it is possible that variations in intelligence may play a significant part in accident susceptibility. Many studies bear out these conclusions. Except at the extremes, intelligence is not associated with accidents to any significant degree; therefore, intelligence tests will not help us predict accidents.

Personality

Some test instruments that measure degrees of emotional reactions have shown a relationship between certain aspects of emotionality and accident frequency.

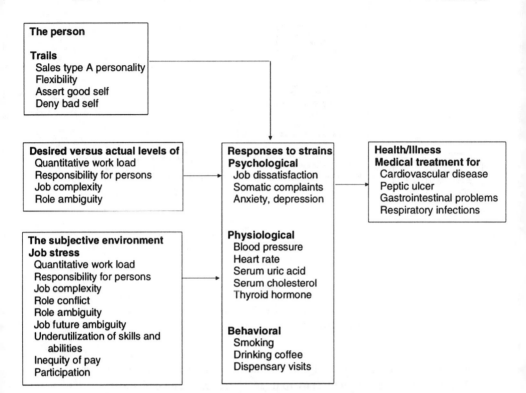

Figure 9-3. Causal relationships regarding worker health. (Adapted from NIOSH, *Job Demands and Worker Health,* HEW Publication No. 75-160. Washington, DC: U.S. Government Printing Office, 1975.)

Even more direct relationships between emotionality and accident occurrence are found when an individual's responses are measured under disturbing and distracting conditions.

Moods also seem to be important. In one study, it was found that half of 400 minor accidents occurred while the employees involved were emotionally low, although this emotional condition existed only 20% of the time. Production was 8% higher during employees' happy moods, showing that emotional conditions favorable to accident prevention are also favorable to production. Also, a worker's emotional adjustment as such is a factor in accidents. Although personality questionnaires have been found inadequate for detecting accident susceptibility, some researchers believe that accidents and poor emotional adjustment are related. And although poor emotional adjustment is related to accident causation, we cannot yet use this knowledge effectively to predict accidents or to select employees, since, in most cases, it is impractical to spend the kind of time and money needed to obtain a psychoanalytic analysis for each applicant.

Capacity

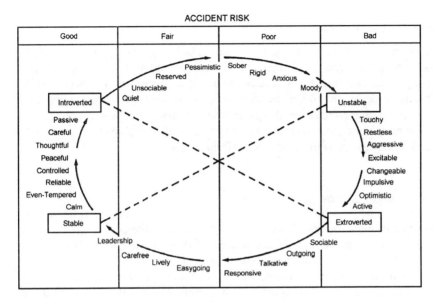

Figure 9-4. The relation of personality to accident risk. (From L. Shaw and H. Sichel, *Accident Proneness*. Oxford, Eng.: Pergamon Press, 1971.)

In one study, a high-accident group was compared with a low-accident group matched for age, education, intelligence, sex, socioeconomic background, and exposure to hazard. The responses of the high-accident group to a sentence-completion test indicated that this group was significantly different from the low-accident group. The low-accident group made more responses indicating optimism, trust, and concern for others.

Mechanical Ability

Several studies made comparing psychological test scores with safety statistics have found no correlation except between test responses that indicate mechanical interest and the accident record. Generally, individuals who have had fewer accidents possess a higher degree of mechanical interest than those who have had a record of frequent accidents. This does not necessarily suggest that tests of mechanical interest should be used as predictors; there is not enough conclusive research on this subject.

In *Accident Proneness,* the classic book by Shaw and Sichel, the results of experimental and statistical studies of personality structure and the relationship of personality to accident risk are discussed. This is summarized by the diagram in Figure 9-4, which suggests a definite relationship between personality type and accident susceptibility.

Knowledge and Skill

Another reason for a low capacity is a simple lack of the necessary knowledge or skill to perform the job. Earlier, we downplayed training as a safety-program component. That does not mean that training is not necessary. It is necessary, but it only achieves one thing: the transmitting of needed knowledge and skills. It does not ensure effective on-the-job behavior. Behavior (error-free performance) is a function of a great many things, one of the many being an employee who has the knowledge and skills needed to perform the job.

Temporary Capacity Reducers

Four items might be considered temporary reducers of job-performance capacity: drugs, alcohol, pressure, and fatigue. Each of these is a major problem in industry today, and each is a definite cause of accidents and injuries on the job.

While we do not have accurate statistics on the drug problem in industry, we do know it exists and that is a major problem. The early recognition of a drug problem seems to be the key to controlling it.

The National Council on Alcoholism recently estimated that 5.3% of the population was addicted to alcohol. Alcohol abuse is a major problem for industry and a major cause of industrial accidents.

Pressure and fatigue also play a major role in reducing work capacity.

While the amount of work and length of working time are factors, other items may be considerably more important. The subjective environment (job stress) undoubtedly causes more pressure and results in more fatigue than anything else. Included in the subjective environment are the complexity of the job, role conflict, role ambiguity, ambiguity of the future of the job, underutilization of a person's skills, and more.

DEALING WITH CAPACITY

There are only three possible managerial approaches to dealing with capacity: (1) proper selection and placement of the employee, (2) changing the employee's behavior, and (3) improving the relationship between the employee and the boss.

Selection and Placement

Control of employee selection and placement has, in recent years, become in-

Capacity   *137*

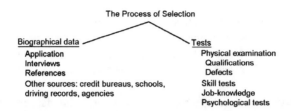

Figure 9-5. Evaluation of potential employees. (From D. Petersen, *Techniques of Safety Management,* 3rd edition, Goshen, NY: Aloray Inc., 1989.

creasingly difficult. Government legislation has increasingly dictated the process, and some industries have almost abdicated control of selection and placement. Even if we cannot fully ensure good selection processes, we certainly ought not to abdicate control in this area entirely. It is far too important to give up. For example, even in the most difficult situations, management can check with previous employers to find out something about an applicant's history.

Figure 9-5 depicts the selection process available to management. Ideally, the process is based on job standards that state that for a particular job a particular type of worker is required. The exactly right person for the job is seldom (if ever) found, but such a goal is still worth striving for. As the figure shows, we have two basic sources of information about an applicant: (1) biographical data and (2) test results. Biographical data are by far the more important source. A person's past performance is still the best indicator of future performance. Tests are of secondary importance (with the possible exception of the physical examination in certain industries). They are important, but should be used to supplement biographical information.

As we have said previously, the various psychological and sensorimotor tests available are not too useful for predicting accident-producing behavior. Depending on the job we are hiring for, the job-knowledge and skill tests might be of value. For instance, if we are hiring a driver, we want to know if the applicant can drive, and drive skillfully. In other situations, a physical examination is most important. The importance and the content of the physical examination should be dictated by the job for which the applicant is being considered and by the physician.

Interviews

The most used selection device in industry today, as in the past, is the interview. Early research by Spreigel and James revealed that 99% of some 852 firms surveyed used the interview. Few people are ever hired who have not been interviewed. Surprisingly, however, there is a preponderance of evidence indicating a general lack of validity for the interview as a selection device. One researcher, while admitting the interview is extremely popular, questions its value.

He asserts that decades-old studies, as well as more recent works, conclude that the interview, as normally conducted in a selection situation, is of little value.

Other research has left us with some information about the interview process as a selective device. In research at McGill University, Webster found that:

1. Interviewers develop a stereotype of a good candidate and seek to match interviewees with stereotypes.
2. Biases are established by interviewers early in the interview and tend to be followed by favorable and unfavorable decisions.
3. Unfavorable information influences interviewers more than favorable information.
4. Interviewers seek data to support or deny hypotheses and, when satisfied, turn their attention elsewhere.

In other words, we say we interview to get factual data on the person, but in actuality we form a very early opinion, make up our minds, and then spend the remainder of the interview time ferreting out facts that will substantiate that early opinion. Thus even the trusted interview may not give us the information we need to hire employees with safe work habits.

Reference Checks

The accident record of a prospective employee can be checked through the previous employer and, in the case of the driving record, through motor vehicle reports. Several researchers question the real value of the information obtained in reference checks, and most agree that if previous employers are contacted, the contact should be by telephone and not by letter. The validity of reference information is highest if the prospective and previous employer know each other.

Selection and Placement as a Safety-Control Process

Selection and placement of employees is a rather weak control device for reducing human error. Getting solid biographical data seems to be our best bet, but good data are difficult to obtain. The application, the interview, and the reference check do not always give us good data, and these are the best devices we have. Tests can be quite valid in certain instances, but in other instances they can be quite invalid—and, in most cases, tests are illegal under EEO guidelines unless statistically validated to job success.

The selection process is also affected by how many applicants there are, current economic conditions, the company's wage rates in comparison to others in the community, and so on. At best, employee selection does not have much effect on human-error reduction.

Behavioral Change

If, as it appears, selection does not help us to reduce human error in the workplace, we end up with error-making employees on our payrolls and in our plants. We then must deal with them as best we can by attempting to change their job behavior so that it will be less error-prone. As shown in Figure 9-6, there are three ways by which we can change the behavior of workers: (1) direct confrontation, (2) training or coaching, and (3) behavior modification.

The workers can respond to direct confrontation in three ways: (1) they can try and succeed; (2) they can try but fail; or (3) they can have an aversive reaction to the whole process, which might take the form of withdrawal, avoidance, or attack. Either of the latter two responses is more likely than the first.

Workers can respond to training or coaching in the same three ways, but the likelihood of trying and succeeding is much higher than with direct confrontation. With behavior modification, the response is unconscious (subconscious), and the result will be success.

Training and Coaching

A well-done training program starts with the development of training objectives. Next, training content and criteria, methods, and materials are developed. After the training program is presented, the results are compared to the criteria. Many, perhaps most, safety training programs are not well done, and these steps are not followed, lessening the likelihood of good results. In simple terms, the training process has three steps:

1. Finding out where we are.
2. Finding out where we want to be.
3. Providing the difference.

The bulk of the training time should be spent on steps 1 and 2. Too often, safety programs are constructed because of a directive from management or because the safety director has jumped to the conclusion that training is needed. Too often, there is no attempt to find out what employees' current knowledge is or even what it is that employees need to know to do their jobs. Usually the training specialist or safety director teaches (or teaches again) whatever it is he or she knows best, whatever is covered in a film or programmed text, or whatever curriculum materials are available. Training is only effective in the reduction of human error when it fills a knowledge gap.

Behavior Modification

As indicated in Figure 9-6, behavior modification is a better behavior-changing device than either direct confrontation or training. It does, however, have some drawbacks.

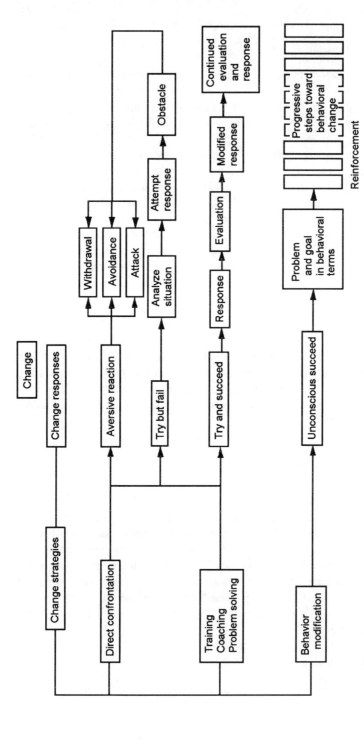

Figure 9-6. Changing on-the-job behavior. (From J. Reitzesk and R. McKenna, *Experiences in Training: The Escondido Papers.* Escondido, CA: Peace Corps, 1971.)

Capacity

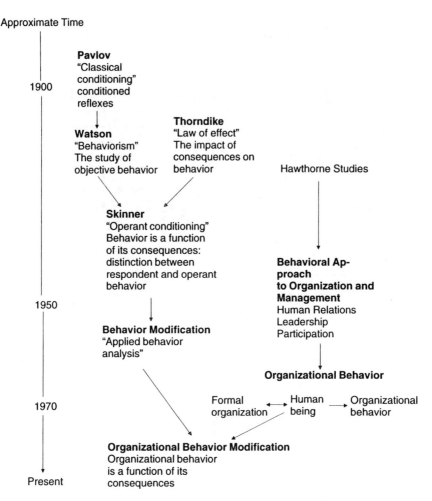

Figure 9-7. Historical development of organizational behavior modification. (From F. Luthans and R. Kreitener, *Organizational Behavior Modification.* Glenview, IL: Scott, Foresman, 1975.)

Behavior modification is a specific technique based originally on the theories and teachings of B. F. Skinner. Skinner's teachings state that behavior is a function of its consequences. If a response is followed by a favorable situation (positive reinforcement), it is likely that that response will be repeated. The historical development of behavior modification in organizations is shown in Figure 9-7, and a model of how behavior modification is used in organizations is shown in Figure 9-8.

The advantages of behavior modification are that it works better than the other devices we use, it teaches the response better, and it is less likely to be forgotten.

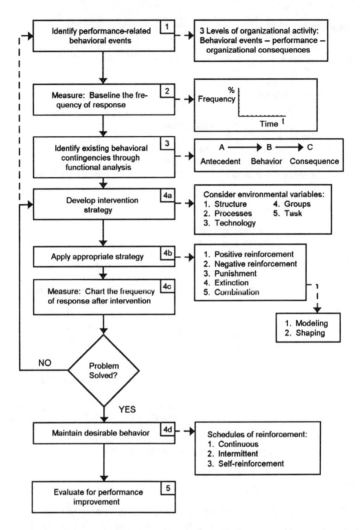

Figure 9-8. Behavior contingency management. (From F. Luthans and R. Kreitener, *Organizational Behavior Modification,* Glenview, IL: Scott, Foresman, 1975.)

It requires that the manager (or whoever is shaping the worker's behavior) decide precisely what the desired behavior is (what is important). One disadvantage of behavior modification has to do with ethical considerations: we change the worker's behavior without his or her knowledge or permission. Another disadvantage is that it takes some time for a supervisor to accomplish the change. Behavior modification is now being used in many areas in industry, but it is not yet being used to any extent in safety situations.

Interpersonal Relations

The interpersonal relationship most relevant to capacity is the worker's relationship with his or her boss. This will be covered in much more detail in later chapters, as it is important to many aspects of human-error reduction. For maximum worker capacity, the boss must perform accurate job analysis prior to worker selection; observe regularly the worker's physical condition and state of mind; impart to the worker necessary knowledge and skills; identify the temporary reduction of worker capacity due to drugs, alcohol, pressure, or fatigue; and administer behavior modification programs. In short, the boss's awareness of the needs and condition of the worker is of crucial importance.

REFERENCES

Deatherage, B. Auditory and other sensory forms of information presentation. In H. Van Cott and R. Kinkade, *Human Engineering Guide to Equipment Design.* Washington, DC: U.S. Government Printing Office, 1972.

Eastman Kodak Company, *Ergonomic Design for People at Work,* Vol. 1 & 2, New York: Van Nostrand Reinhold, 1986.

Hammer, W. *Occupational Safety Management and Engineering.* Englewood Cliffs. NJ: Prentice-Hall, 1976.

Kantowitz, B. H. and R. D. Sorkin, *Human Factors: Understanding People-System Relationships,* New York: John Wiley and Sons, 1983.

Luthans, F. and R. Kreitener. *Organizational Behavior Modification.* Glenview, IL: Scott, Foresman, 1975.

McCormick, E. *Human Factors in Engineering and Design.* New York: McGraw-Hill, 1976.

McFarland, R. In J. Recht, *Systems Safety Analysis: Error Rates and Cross.* Chicago: National Safety Council, 1970 Reprint.

MIL-STD1472C, *Military Standard, Human Engineering Design Criteria for Military Systems, Equipment and Facilities,* U.S. Department of Defense, Washington, DC, May 2, 1981.

NIOSH. *Job Demands and Worker Health.* HEW Publication No. 75-160. Washington: U.S. Government Printing Office, 1975.

Petersen, D. *Techniques of Safety Management,* Goshen, NY: Aloray Inc., 1989.

Reitzesk, J. and R. McKenna. *Experiences in Training: the Escondido Papers.* Escondido, CA: Peace Corps, 1971.

Rockwell, T. In J. Surry, *Industrial Accident Research.* Toronto: University of Toronto, 1968.

Salvendy, G., *Handbook of Human Factors,* New York, John Wiley and Sons, 1987.

Sanders, M. S. and E. J. McCormick, *Human Factors in Engineering and Design,* Hightstown, NJ, McGraw-Hill Book Company, 1987.

Shaw, L. and H. Sichel, *Accident Proneness.* Oxford, Eng.: Pergamon Press, 1971.

Silvern, L. *Systems Engineering of Education.* Houston: Education and Training Consultants, 1971.

Sinaiko, H. Selected papers on human factors in the design and use of control systems. Chicago: National Safety Council, 1970 Reprint.

Spreigel, W. and J. James. In O. Wright, Summary of research on the selection interview since 1964. *Personnel Psychology* 22:391–413, 1969.

Van Cott, H. and R. Kinkade. *Human Engineering Guide to Equipment Design.* Washington, DC: U.S. Government Printing Office, 1972.

Webster, J. In O. Wright, Summary of research on the selection interview since 1964. *Personnel Psychology* 22:391–413, 1969.

Wood, F. In G. Lippitt, *Visualizing Change.* La Jolla, CA: University Associates, 1976.

Woodson, W. and D. Conover. *Human Engineering Guide for Equipment Designers.* Berkeley: University of California Press, 1966.

Woodward, J. L. (ed.), *Proceedings of the International Symposium on Preventing Major Chemical Accidents,* Center for Chemical Process Safety of the American Institute of Chemical Engineers, Washington, DC, February 3–5, 1987.

Chapter *10*

Load

Load is a catch-all term that refers to the total amount of work and stress placed on the worker from all sources, at work and at home. Thus load means not only the physical, physiological, and psychological load that is part of the job, but also the load the worker carries as a result of home worries and stresses.

In this chapter, we rather arbitrarily classify load into two categories, short-term and long-term. By short-term we mean the present work situation: what the worker is facing right now on the job. Short-term load might be different tomorrow, might be heavier or lighter even a few minutes from now, and is pretty much a function of current external influences and internal feelings generated by the current work situation. Long-term load tends to be the stresses the worker has due to life situation or current mental health. If a worker is in a particularly difficult home situation, for example, due to a death in the family or marital difficulties, he or she carries some rather long-term stresses that must constantly be coped with. Examples of both short- and long-term stresses are shown in Figure 10-1 in two categories, physiological and psychological, along with measures of strain resulting from those stresses.

SHORT-TERM LOAD

Some aspects of short-term load are the current task the worker must do (physical work, information processing, etc.), the environmental load (the situation in which the worker must perform that task), perhaps a circadian load, perhaps a load due to either fatigue or boredom, and various kinds of stresses employers

	Work	Heavy work Immobilization	Information overload Vigilance
Sources of stress	Environment	Atmospheric noise/vibration heat/cold	Danger Confinement
	Circadian habits	Sleep loss	Sleep loss
		Physiological	Psychological
	Chemical	Electrical	Physical	Activity	Attitudes
Measures of strain	Blood content Urine content Oxygen consumption Oxygen deficit Oxygen recovery curves Calories	EEG EKG EMG EOG GSR	Blood pressure Heart rate Sinus arrhythmia Pulse volume Pulse deficit Body temperature Respiratory rate	Work rate Errors Blink rate	Boredom Other attitudinal factors

Figure 10-1. Primary sources of stress, and strain induced by stress. (Adapted from W. Singleton, The measurement of man at work with particular reference to arousal, in *Measurement of Man at Work,* edited by W. Singleton, J. Fox, and D. Whitfield. London: Taylor and Francis, 1971.)

might put on the worker, including, perhaps, the additional stress of having to perform tasks in a dangerous or hazardous work situation.

The Task Itself

What does the worker have to do in comparison to what he or she is able to do? First the physical task: How much does the worker have to lift, and in what position? Can the worker reach what he or she needs to? Is the size of the object, the force necessary to move it, and so on, within the worker's capabilities? Analysis of the physical task is simple and is certainly within the capabilities of any line manager. These physical considerations are usual and normal parts of job-safety analyses. They are often included in job descriptions. The identification of physical task requirements and the comparison of those requirements to worker capabilities is, and has been, part of safety technology for years.

The physical aspects of the task are only a small part of the task analysis. In addition to physical considerations, mental and psychological considerations of the task must be evaluated. Mental load might include whether or not the information-processing aspects of the task exceed the worker's capabilities. Psychological load might include task ambiguity, task-success criteria, feedback, task confusions, and shifting goals and tasks. Figure 10-2 identifies some of the effects of the task on the stress load of the worker. If the task is physically too difficult, if the criteria for success are unclear, if the responses required from the worker are constantly

Load

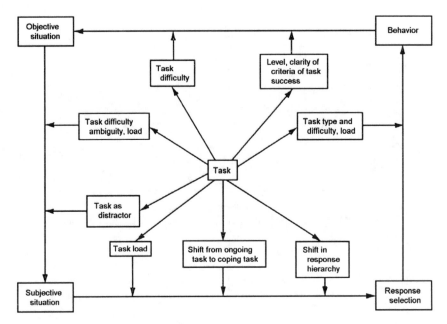

Figure 10-2. Effects of the task on various stages of the stress cycle. (Adapted from J. McGrath, Stress and behavior in organizations, in *Handbook of Industrial and Organizational Psychology,* edited by M. Dunnette. Chicago: Rand-McNally, 1976.)

changing, if the task itself is constantly changing, if there are constant distractions in the performance of the task, and if the task itself is ambiguous, stress is placed on the worker.

Environmental Load

A second load factor is environmental load. This refers not to the work itself and its effect, but rather to the environment in which the work is done. It is fairly well documented that comfort, efficiency, and safety are all influenced by the environment in which the work takes place. Figure 10-3 shows the comfort ranges and tolerance levels of human beings for various environmental variables. When these variables are within the comfort zone, both efficiency and safety are improved.

There are numerous examples of this. Broadbent, for example, investigated the relationship between environmental noise and error rates and found that, under continuous, 100-db noise conditions, performers of tasks requiring vigilance made more errors than workers performing under quieter conditions. Jerison reported that under noisy conditions workers took longer to perform tasks and that noise affected

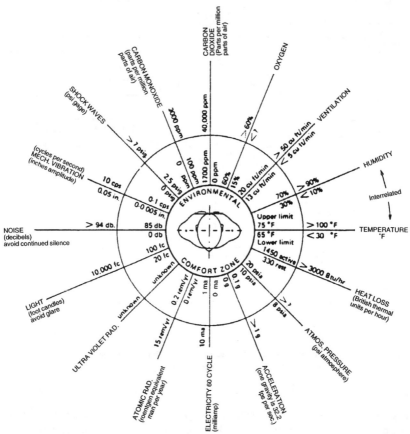

Figure 10-3. Physical environmental variables that influence driver comfort, efficiency, and safety. (Adapted from R. McFarland, Application of human factors engineering to safety engineering problems, in J. Widener, Ed., *Selected Readings in Safety.* Macon, GA: Academy Press, 1973.)

the workers' ability to estimate the length of time the tasks would take. There are many experiments about the relation of stress to efficiency, all of which seem clearly to indicate a strong negative relationship. For instance, the effects of high temperature and humidity on performance have been studied (Figure 10-4).

Table 10-1 indicates the relation of physical environmental stress (force) to injuries. Traditionally, safety professionals have concentrated on identifying the relationship of various environmental factors to actual resulting injury or illness. Perhaps even more important is the relation of these environmental factors to apparently nonrelated accidents and injuries. Environmental stresses can be indirect causal factors in many injuries. For example, when the comfort ranges depicted in Figure 10-3 are exceeded, there is an increased probability that accidents will

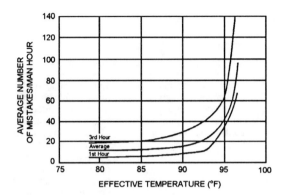

Figure 10-4. Effect of high temperature and humidity on performance of tasks requiring skill. (Adapted from R. McFarland. Application of human factors engineering to safety engineering problems, in J. Widener, Ed., *Selected Readings in Safety*. Macon, GA: Academy Press, 1973.)

occur. Other studies indicate that under stressful conditions even powerful incentives fail to change the situation. For instance, one researcher found that powerful incentives failed to prevent the deterioration of physical output when workers were subjected to the stress of high humidity and temperature.

Fatigue and Boredom

Fatigue is the product of chemical and physical changes in the nerves and muscles of the body that make it difficult to continue to work. Boredom is perhaps a more serious problem in industry today. Boredom sets in when the worker loses interest in the job, and time hangs heavy on his or her hands. Boredom and fatigue are closely related; boredom can actually cause fatigue. In any event, both are stresses. It is possible to compare the effects of boredom and fatigue by looking at production curves for each. Figure 10-5 shows a typical fatigue chart. The curve shows the performance of highly motivated workers doing hard work. After a warm-up period, production drops steadily until lunch time. After lunch, it rises for a short period and then drops steadily until the end of the day. A boredom curve gives a different picture (Figure 10-6). Dips show up in the middle of the morning and afternoon, but there are substantial upswings at the ends of these periods. While fatigue is caused by the excessive use of the nerves and muscles of the body, boredom is caused by a combination of many factors, primarily (1) lack of variety, (2) lack of opportunity, (3) inadequate sense of accomplishment, (4) inability to control one's own work pace, and (5) the need for only superficial attention or none at all. Boredom seems to be a much bigger factor in accident causation than fatigue.

Table 10-1. Injuries caused by delivery of energy in excess of local or whole-body injury thresholds

Type of Energy Delivered	Primary Injury Produced	Examples and Comments
Mechanical	Displacement, tearing, breaking and crushing: predominanatly at tissue and organ levels of body organization	Injuries resulting from the impact of moving objects such as bullets, hypodermic needles, knives, and falling objects; and from the impact of the moving body with relatively stationary structures, as in falls, plane and auto crashes. The specific result depends on the location and manner in which the resultant forces are exerted. The majority of injuries are in this group.
Thermal	Inflammation, coagulation, charring, and incineration at all levels of body organization	First, second, and third-degree burns. The specific result depends on the location and manner in which the energy is dissipated.
Electrical	Interface with neuromuscular function, and coagulation, charring, and incineration, at all levels of body organization	Electrocution, burns. Interference with neural function, as in electroshock therapy. The specific result depends on the location and manner in which the energy is dissipated.
Ionizing radiation	Disruption of cellular and subcellular components and function	Reactor accidents, therapeutic and diagnostic irradiation, misuse of isotopes, effects of fallouts. The specific result depends on the location and manner in which the energy is dissipated.
Chemical	Generally specific for each substance or group	Includes injuries due to animal and plant toxins, chemical burns, as from KOH, Br_2, F_2, and H_2SO_4 and the less gross and highly varied injuries produced by most elements and compounds when given in sufficient dose.

Source: Adapted from R. McFarland, Application fo human factor engineering to safety engineering problems, in J. Widener, Ed., *Selected Readings in Safety.* Macon, GA: Academy Press, 1973.

Shift Work

Much has been investigated and written about the relationship of various forms of shift work to human error and accidents. Glenn McBride and Peggy Westfall, in their manual *Shiftwork: Safety and Performance,* point out some of the relationships:[1]

---

1. Reprinted with permission from G. McBride and P. Westfall, *Shiftwork: Safety and Performance,* Chicago, National Safety Council, 1995.

Load   *151*

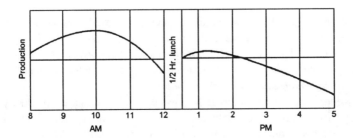

Figure 10-5. Fatigue chart. (From George Strauss, Leonard R. Sayles, *Personnel,* © 1986, pp. 34, 35. Reprinted by permission of Prentice-Hall, Inc., Englewood Cliffs, New Jersey.)

Chernobyl is a well-known shiftwork catastrophe that occurred in the wee morning hours. This case of *immediate human error coupled with planning and engineering Set-Ups* continues to spin off disastrous consequences. Recently, the Supreme Soviet legislature adopted a $26 billion program to move 200,000 additional residents from the contaminated area. Officially, there were 31 fatalities. However, unofficial reports estimate a minimum of 250 deaths. The spread of Iodine-131 quickly collected in the thyroids of thousands of radiation victims.

Other late night incidents often associated with shiftwork include the Bhopal Gas Release, the Three-Mile Island nuclear incident, and the Exxon Valdez oil spill.

Legislatures have gotten into the act. In November, 1991, the National Commission on Sleep Disorder Research, mandated by Congress, completed their 20-month study on inadequate sleep in the United States.

The commission concluded that insufficient sleep is an unseen killer. Merrill Mitler reports that millions of accidents each year are caused by those drivers and workers who try to accomplish too much with too little sleep.

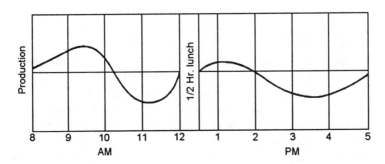

Figure 10-6. Boredom chart. (From George Strauss, Leonard R. Sayles, *Personnel,* © 1986, pp. 34, 35. Reprinted by permission of Prentice-Hall, Inc., Englewood Cliffs, New Jersey.)

Mitler writes that these findings quickly followed a September 1991 report by the Congressional Office of Technology Assessment. This report said that the *"biological cycles that affect sleep, fatigue and human error in the workplace have been underestimated or ignored by federal regulators, industrial management, and labor leaders."*

Mitler says that recent research shows that individuals are more likely to make errors between midnight and 6 a.m. even when they have slept seven or eight hours during the day. *(And they are EVEN MORE LIKELY to commit errors when they have not slept well during the previous 24 hours.)* Gene Lowe of Dow Chemical labels the dangerous hours as "the Danger Zones." There are actually two Danger Zones each 24 hours.

The most ominous Danger Zone, that Mitler described, occurs nightly after midnight.

The Danger Zone "alertness dips" follow the daily body temperature cycle each 24 hours. This "circadian cycle" is usually aligned with real time (even for many long term "night workers").

Also, an incredible percentage of *major catastrophes* have occurred around the Danger Zone time period! The Exxon Valdez touched ground just a few minutes after midnight. This is the same hour of the major gas release by the Union Carbide Plant in Bhopal.

The Danger Zones have been verified by many expert studies, including Merrill Mitler's. Researcher Michael Coplen also studies this phenomenon. He cites a study of train engineers in Sweden: 10 percent reported micro-sleep Slips (of 30 seconds to 3 minutes) *nearly every night.* In the same article, he reports that a Japanese study of 198 near-miss train accidents reported that *82 percent occurred between midnight and 8 a.m.*

Shiftwork Set-Ups have led to many train incidents over the years. A sleepy engineer and crew were the probable cause of the 1988 head-to-head collision of two Conrail freight trains near Thompsontown, Pennsylvania according to NTSB investigators. This crash cost four lives and $6 million.

A Burlington Northern freight train struck another Burlington train head-on at 3:55 a.m. and five crewmembers were killed. *Nine days later, another Burlington rear-ended another Burlington around 4:18 a.m.* near Pedro, Wyoming and two were killed immediately. The National Transportation Safety Board investigated the accidents and determined that the employees' falling asleep (Attention Slips) was a probable cause in both incidents.

The U.S. Dept of Transportation reports that up to 200,000 traffic accidents each year may be sleep-related. They claim that 20 percent of all drivers have dozed off (micro-sleep Slips) at least once while behind the wheel. Time Magazine says that the truck driver portion is particularly vulnerable. A long-haul driver covering up to 4,000 miles in seven to 10 days often averages only two to four hours of sleep a night.

The Better Sleep Council has become concerned about the dozing statistics. They tell Americans that *falling asleep at the wheel is second only to alcohol as a cause of U.S. accidents.* They mention that drowsy vehicle drivers (including trucks, automobiles, etc.) kill as many as 6,500 per year.

A study of 182 fatal truck incidents in 8 states between October 1987 and September 1988 supports this. The National Transportation Safety Board found that *impairment of the driver due to driver fatigue was a primary cause* in 31.5% of the crashes.

In a human factors analysis of the Challenger incident, they mention that excessive overtime and fatigue of *key individual personnel affects the safety of the entire operation. This is more important than the simple overtime average of the group!* During the month of the disaster, several key individuals were essentially working two shifts. From this, we have learned an important lesson: watch the overtime of key personnel! This type of data will certainly be identified more and more during incident investigations.

Then, in a linear fashion, we can predict the details of some important (not too distant) future headlines for our own companies and others. We can predict that a statistically high number of headline incidents will occur (or be set up!) after midnight due to human error and these incidents will involve one or more employees with under two years experience. Many of the incidents will involve Set-Ups such as poor fitness and health, overwork, and inappropriate meals. These will have led to fatigue, jet lag, and slow response time—and will have caused a Slip. Additionally in many industries, the slips committed during start-ups, shutdowns and shift changes by newly-formed crews will make headlines!

Anxiety and Other Distracting Stresses

Anxiety, fear, and similar distracting stresses have been studied only a little in terms of their relation to performance. Fear seems to improve performance when it is moderate and temporary. It interferes with performance, however, when it reaches too high a level or when it is sustained over a long period of time. For the most part, it appears that experience on the job (confidence) is the key to controlling job-related anxiety and fear. Mean fear rates of operating personnel in the chemical industry have been computed (Figure 10-7). As the chart shows, the key to the control of job-related anxiety seems to be confidence that comes from experience. What are the causes of anxiety and tension on the job? Handy reports on the major tension-producing factors at various levels of management (Table 10-2).

McGrath, in his studies of stress and behavior in organizations, has described some important relationships, which are charted in Figure 10-8. His research shows that if arousal and stress are constant, the level of performance is inversely related to level of task difficulty. If task difficulty is held constant and the level of arousal

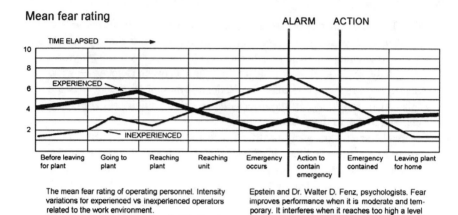

Figure 10-7. Mean fear rating. (From A. Yariger, Stress—the father of disaster, Reprinted with permission from *Industry Week* (Sept. 20, 1971). Copyright Penton Publishing, Inc., Cleveland, Ohio.

is increased, performance improves. He describes stress as a function of increasing difficulty and increasing arousal. The net result, then, of increasing stress (difficulty and arousal) is as shown in the bottom graph in Figure 10-8. Performance increases with increasing stress up to a point, and then it begins to decrease. McGrath proposes the stress cycle shown in Figure 10-9. As a stressful event occurs, people appraise it and decide what to do based on the perceived situation. That decision is the selection of an appropriate response, which, translated into overt action, is behavior. Personal behavior in the situation is the outcome, and it changes the situation, triggering the cycle again.

Stresses from Hazard and Danger

Working in a hazardous or dangerous situation is in itself stressful to the worker. This is perhaps most obvious in wartime in combat situations, but is also the case in most work situations to a lesser degree. Hazardous or dangerous situations increase arousal level and increase the task difficulty, so the model in Figure 10-8 applies. Performance levels increase with danger to a certain degree, then decrease. This is not to say that the hazard level in the workplace should be kept at a medium level. Obviously, over a period of time arousal levels due to danger would not be maintained, and task difficulty would create deterioration in performance over the long term.

Table 10-2. Manager response to tension

Job-Related Tension-Index Items	Percentage of All Managers Indicating Item as a Pressure-Inducing Item	Percentage of Top Managers Only	Percentage of Middle Managers Only	Percentage of Difference
Having to decide things on the job that affect the lives of individuals, people you have come to know	88%	83%	90%	7%
Feeling unable to influence your immediate superior's decisions and actions that affect you	85%	67%	90%	23%
Thinking that you'll not be able to satisfy the conflicting demands of various people over you	81%	67%	85%	18%
Fact that you can't get the necessary information needed to carry out your job	77%	67%	90%	23%
Feeling that you have to do things on the job that are against your better judgment	73%	67%	75%	8%
Feeling that you have too little authority to carry out the responsibilities assigned to you	69%	50%	75%	25%
Feeling that you have too heavy a work load, one you can't possibly finish in an ordinary work day	66%	66%	66%	0%
Being unclear on just what the scope and responsibilities of your job are	65%	33%	75%	42%
Feeling that you may not be liked or accepted by the people you work with	61%	33%	70%	37%
Not knowing just what the people you work with expect of you	58%	34%	65%	31%
Feeling that your job tends to interfere with your family life	50%	50%	70%	20%
Feeling that you're not fully qualified to handle yourself	50%	33%	65%	32%
Not knowing what your supervisor thinks of you or how he evaluates your performance	46%	50%	70%	20%
Not knowing what opportunities for advancement or promotion exist for you	39%	17%	45%	28%
Thinking that the amount of work you do may interfere with how well it gets done	23%	33%	20%	13%

Source: Adapted from C. Handy, *Understanding Organizations.* Middlesex, Eng. Penguin Books, 1976.

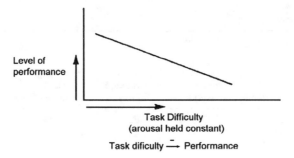

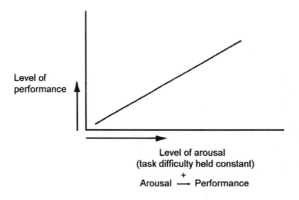

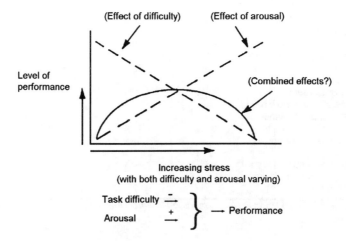

Figure 10-8. Effects of task difficulty, level of arousal, and increasing stress on level of performance. (From J. McGrath. Stress and behavior in organizations, in *Handbook of Industrial and Organizational Psychology,* edited by M. Dunnette. Chicago: Rand-McNally, 1976.)

Load   *157*

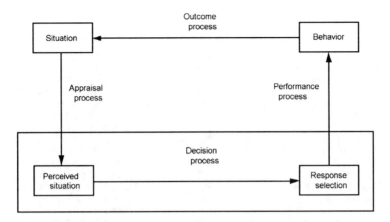

Figure 10-9. The stress cycle. (From J. McGrath. Stress and behavior in organizations, in *Handbook of Industrial and Organizational Psychology,* edited by M. Dunnette. Chicago: Rand-McNally, 1976.)

LONG-TERM LOAD

In recent years, there has been considerable discussion and research on the relationship between a person's life situation and the likelihood that he or she will be involved in accidents. Probably the most important concept in this area is the concept of life change units (LCUs). Several researchers have examined the relationship between life events that have happened to an individual in the recent past and that individual's predisposition in the near future to illness or injury. Table 10-3 summarizes the work of R. A. Alkov, one of the primary researchers evaluating life change units. Using the table, a person can assess his or her current life situation, quantify it, and predict the likelihood of a health change or injury in the near future. While we do not know why the relationship exists, its existence has been well documented. Perhaps the human being is so busy subconsciously coping with these life-situation problems that his or her defenses against illness and injury are down. These personal-situation loads are definitely a factor in injury causation. To quote Dr. Alkov:[2]

> About 25 years ago a psychiatrist, Dr. Thomas H. Holmes, now at the University of Washington, found that many diseases were caused by life events—diseases such as colds, tuberculosis and skin disease. In studies, he asked more than 5,000 patients to tell about life events that preceded their

---

[2]. Reprinted from R. Alkov, The life change unit and accident behavior. *Lifeline,* Norfolk, VA, U.S. Naval Safety Center, 1972.

Table 10-3. Life change units

Rank	Life Event	Mean Value	Rank	Life Event	Mean Value
1	Death of spouse	100	22	Change in work responsibilities	29
2	Divorce	73	23	Son or daughter leaving home	29
3	Marital separation	65	24	Trouble with in-laws	29
4	Jail term	63	25	Outstanding personal achievement	28
5	Death of close family member	63	26	Spouse begins or stops work	26
6	Personal injury or illness	53	27	Begin or end school	26
7	Marriage	50	28	Change in living conditions	25
8	Fired at work	47	29	Revision of personal habits	24
9	Marital reconciliation	45	30	Trouble with boss	23
10	Retirement	45	31	Change in work hours, conditions	20
11	Changes in family member's health	44	32	Change in residence	20
12	Pregnancy	40	33	Change in schools	20
13	Sex difficulties	39	34	Change in recreation	19
14	Gain of new family member	39	35	Change in church activities	19
15	Business readjustment	39	36	Change in social activities	18
16	Change in financial state	38	37	Mortgage or loan under $60,000	17
17	Death of close friend	37	38	Change in sleeping habits	16
18	Change to different line of work	36	39	Change in no. family get-togethers	15
19	Change in no. arguments with spouse	35	40	Change in eating habits	15
20	Mortgage over $60,000	31	41	Vacation	13
21	Foreclosure of mortgage or loan	30	42	Christmas	12
			43	Minor violations of the law	11

Example of events that can occur to a person over a short interval:	Mean Value
Marital separation	65
Change in responsibility at work	29
Change in living conditions	25
Revision of personal habits	24
Change in working hours or conditions	20
Change in residence	20
Change in recreation	19
Change in social activities	18
Change in sleeping habits	16
Change in eating habits	13
Total	249

A pilot study indicated that when a person undergoes various life events over a short interval of time, as in the example shown, the following tendencies appeared:

Values	Percentage of Those Investigated	Results
150–199	37	Health change within 2 years (generally 1)
200–299	51	Health change within 2 years (generally 1)
Over 300	79	Injuries or health changes

Source: R. A. Alkov, the life change unit and accident behavior, *Lifeline.* Norfolk, VA: U.S. Naval Safety Center, September–October 1972.

illnesses. These covered a wide range—death of a spouse, a change of job, divorce, birth of a child, etc. These events were noted on the patient's health records and were referred to in later visits. Only some of the events were negative or stress producing—most were ordinary events of the American way of life: family events, economic events, vacations, retirement, etc. The important factor was change—desirable or undesirable—in ongoing life styles, which would require adaptive or coping behavior.

Commander Richard Rahe, MC, USN, now at the Navy's Medical Neuropsychiatric Research Unit in San Diego and Dr. Holmes found that illness follows a cluster of events that require life adjustment. Life events tended to cluster in a 24-month period before the onset of tuberculosis, heart disease, skin disease, and hernia.

In order to identify those life events that most frequently preceded illness, they asked 394 persons to rate the amount of social readjustment required for each of the 43 events listed most frequently by patients. By an arbitrary method the life event given the top ranking by the judges, death of a spouse, was weighted 100 points on the scale. Using the rank order method the other weights were calculated.

In a pilot study it was found that of those persons who reported Life Change Units (LCU's) that totaled between 150 and 199 points, 37 percent had associated health changes within a two-year period of such life crises. Of those with between 200–299 Life Change Units, 51 percent reported health changes and of those with over 300 LCU's, 79 percent had injuries or illnesses to report. On the average, health changes followed life crises by one year. The knowledge that the emotionally stressed individual may be more prone to illness and accident is not new. It has long been known, for example, that overstressed individuals often engage in irrelevant activities or rigid stereotyped behavior and experience loss of discriminative skill and mental efficiency. The safe performance of complex tasks (such as those demanded in aviation) is improbable in such a psychological context.

Of course, each person is an individual with his own unique personality and method of handling stress. Some people are more susceptible to the effects of emotional factors than others. These changes in an individual's daily style of living and personal family matters may have little influence on his performance until they add up to an unbearable psychological burden. It is incumbent upon those in supervisory positions to monitor and observe the effects of turmoil in the personal lives of their aviators on their performance in flight. If necessary and upon consultation with the medical officer, the aircrew member might be temporarily grounded or provided with leave until his problems are resolved.

The knowledge that changes in one's life style, whether positive or negative, can tax a person's coping ability should enable us to construct a life changes factor weighting score for each individual which will allow us to predict within certain probability limits, the likelihood of his being involved in a human error mishap.

Dealing with Load

There are a number of traditional safety approaches that help to deal with load and that employers should continue to carry out. Job-safety analyses and job-hazard analyses are two such systematic techniques for identifying and removing hazards. Safety professionals have for years used these techniques, plus inspections and other monitoring devices, to increase job safety.

Another traditional task for safety professionals has been to improve the interpersonal skills of supervisors. Worry and stress loads can be controlled best by supervisors who know their people well enough to know when a worker has a problem that needs to be dealt with. Human relations training—and, more recently, transactional analysis—has helped supervisors better understand and help the workers they oversee.

References

Alkov, R. The life change unit and accident behavior. *Lifeline.* Norfolk, VA: U.S. Naval Safety Center, 1972.

Better Sleep Council, P.O. Box 13, Washington, DC.

Broadbent, D. Noise, paced performance and vigilance tasks. *British Journal of Psychology* 44:295–303, 1953.

Bronner, K. Industrial implications of stress. In J. Levi, *Emotional Stress.* New York: American Elsevier, 1961.

Caplan, G. *Support Systems and Community Health.* New York: Behavioral Publications, 1974.

Coplen, M. Does shift work shift risks?, *Safety & Health* June 1988.

Dickens, D. Bored to death. In *Job Safety and Health.* Washington, DC: U.S. Government Printing Office, 1975.

Dodge, D. and W. Martin. *Social Stress and Chronic Illness.* London: Notre Dame, 1970.

Friedman, M. and R. Rosenman. *Type "A" Behavior and Your Heart.* New York: Knopf, 1974.

Graeber, R. Aircraft fatigue and circadian rythmicity, *Human Factors in Aviation,* January 1988.

Handy, C. *Understanding Organizations.* Middlesex, Eng.: Penguin Books, 1976.
Jerison, H. Differential effects of noise and fatigue on a complex counting task. WADC Technical Report No. 55-359, 1956.
Holmes, T. and M. Masuda. Psychosomatic syndrome. *Psychology Today,* April 1972.
Holmes, T. and T. Holmes. Short term intrusions into the life style routine. *Journal of Psychosomatic Research* 14(2):121, 1970.
McBride, G. and P. Westfall, *Shiftwork: Safety and Performance.* Chicago: National Safety Council, 1995.
McFarland, R. Application of human factors engineering to safety engineering problems. In *Selected Readings in Safety.* Macon, GA: Academy Press, 1973.
McGrath, J. Stress and behavior in organizations. In *Handbook of Industrial and Organizational Psychology,* edited by M. Dunnette. Chicago: Rand-McNally, 1976.
Mitler, M. Tired workers cause millions of accidents a year. *New York Times,* January 11, 1992.
Rahe, R. Life change and subsequent illness reports. In E. Gunderson and R. Rahe, *Life Stress and Illness.* Springfield, IL: Thomas, 1974.
Reamer, F. Life change and stress—propensity to accident and illness. Unpublished master's thesis, University of Arizona, Tucson, 1971.
Rigby, L. and A. Swain. Effects of assembly error on product acceptability and reliability. In *Proceedings* of the 7th Annual Reliability and Maintainability Conference, American Society of Mechanical Engineers, New York, July 1968.
Rook, L. *Reduction of Human Error in Industrial Production.* SCTM-93-62. Albuquerque, NM: Sandia Labs, 1962.
Schlesinger, L. Can you die of boredom? *Professional Safety,* September 1974.
Singleton, W. The measurement of man at work with particular reference to arousal. In *Measurement of Man at Work,* edited by W. Singleton, J. Fox, and D. Whitfield. London: Taylor and Francis, 1971.
Strauss, G. and L. Sayles. *Personnel—The Human Problems of Management.* Englewood Cliffs, NJ: Prentice-Hall, 1960.
Swain, A. *A Method for Performing a Human Factors Reliability Analysis.* ACR-685. Albuquerque, NM: Sandia Labs, 1963.
Swain, A. *THERP.* SC-R-64-1338. Albuquerque, NM: Sandia Labs, 1964.
Swain, A. A work situation approach to improving job safety. In *Selected Readings in Safety.* Macon, GA: Academy Press, 1973.
Yariger, A. Stress—the father of disaster. *Industry Week,* September 20, 1971.

Chapter *11*

State

The concept of state includes the individual's motivational state, attitudinal state, arousal state, and perhaps, if we believe in the concept, biorhythmic state. In this chapter, we examine each of these aspects of state, and take a look at what managers can do about them. In some situations, a person's state can be influenced; in others, it can be affected very little. We ought, however, at least to be able to recognize a worker's state and exert additional influence where we can.

MOTIVATIONAL STATE

Probably the biggest single area to examine is the motivational state of the employee, and perhaps this area is the most complex. The motivational field of the employee is a function of many things. Figure 11-1 shows the determinants of employee performance. This model suggests that performance at the employee level is a function of two things: whether or not the employee is able to do the task, and whether or not he or she wants (is motivated) to do it. Ability is a function of selection (Do we have the right person for the task?) and of training (Have we taught the worker how to do it?). Motivation is a function of the organizational climate, the style of the boss (and his or her interpersonal relationship with the worker), the job itself and its motivational factors, the worker's personality, the peer group, and perhaps the union. We look at each of these influences in this chapter and again in later chapters, as some of them are behind the worker's decision to err as well as the state of the worker.

163

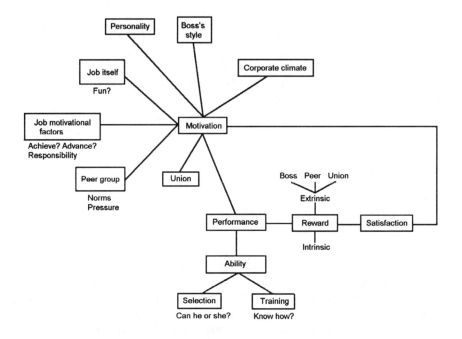

Figure 11-1. Motivational factors influencing employee performance.

Climate

There seems little question that the way the worker sees the company safety program strongly influences not only behavior on the job but ability to learn from, and to respond to, safety materials. Workers characterize the safety programs where they work quite consistently within any given company. This was brought out by Social Research, Inc., in a study in which the researchers found four general company types:

1. The overzealous company. This kind of company requires that a great amount of safety equipment be worn, and that machines be so guarded as to make them difficult to get near and work with. The tone of the safety program is heavy. Such a company is likely to impose harsh punishment (a three-day layoff, for example) for some minor infraction of safety procedures. There seems to be an excess of meetings, films, manuals, and preachments about safety, which often do not involve workers directly in any personal way. In such companies, response to safety-education materials is not lively, and workers feel overexposed to the subject.

2. The rewarding company. This company type might offer prizes for safety records or for entering into safety-slogan competitions. While the prizes are relatively small, there is a sense of competition for them. Employees in such companies feel that safety programs are important, and that the company is acting responsibly about safety matters.
3. The lively company. Such a company has a safety program that stimulates competition among the various plants, that offers plaques, that has scoreboards to record the number of hours passed without accident, or that posts at the plant entrance a sign proclaiming its continuing safety record. The lively companies are companies that teach the workers to identify with safety goals, and in which, as a result, the employees are proud of their record. Safety in these companies becomes one of the lively aspects of the job and is more than avoiding risk or accident; it is a concrete symbol and goal.
4. The negligent company. This is the kind of company that seems to have safety programs only after a major accident happens. The workers here feel that the company does not really care, and that it passes out safety equipment and information because it is the current custom and because the company wishes to protect itself.

Social Research, Inc. further reported that the friendliness of safety professionals, the manner in which safety is talked about and promoted, and the efficiency with which matters are attended to vary among the four types of companies listed. In the overzealous company, the safety professionals usually seem loud and harsh, overly watchful, and too quick to criticize. In the rewarding and lively companies, safety people are usually seen as friends, as people much like the workers. In the more lax organizations, the workers complain that there are not enough safety officials around, and that they doubt that reporting to safety officials does much good, since little happens. These conclusions were reached after detailed, in-depth interviews of workers, interviews designed to assess the need for and worth of safety media. This study effectively identified the types of climates common in industrial safety programs.

Safety-program climate is only one aspect of total corporate climate. General corporate climate is a major influence on the behavior of both managers and employees. According to Burt Scanlon of Oklahoma University, one point of concern is that employees may not perceive the organization's climate and philosophy as management intends them to. Two possible reasons may contribute to this difference between desired and actual perception. First, perhaps not enough effort has been expended in communicating the guiding philosophy down the line. Second, there may be a discrepancy between what is professed and actual occurrences. The individual's closest point of contact with the organization is his or her immediate superior. If the superior's actions do not reflect the organizational philosophy, a perception discrepancy occurs.

According to Scanlon, some of the climate requirements basic to maximum individual performance in an organization are as follows:

1. There must be central, overall goals or objectives toward which the organization is striving.
2. The objectives must be communicated down the line with the idea of getting commitment as to value, reasonableness, feasibility, and so on.
3. Functional areas, departmental units, and individuals must also have specific goals to obtain. These must be derived from the central goals, and interrelationships must be perceived.
4. The interdependence of all subunits within the organization in the accomplishment of results must be clearly established, and a framework for interunit cooperation must exist.
5. Meaningful participation on the part of the individual should be the keynote. This means participation in the sense that the individual has a real part in determining his or her job objectives.
6. There must be freedom to work, in the sense that the worker has an opportunity to control and adjust his or her own performance without first being exposed to authority and pressure from a superior. In addition, workers must receive support and coaching from their superiors.

M. Scott Myers specified a number of specific climate factors as necessary before motivation can occur:

1. Expansion. In a growing organization, it is hard to be a bad manager simply because as the organization expands it offers increasing opportunities to achieve, to grow personally and professionally, and to acquire a sense of responsibility.
2. Delegation. Delegation can be encouraged by the manner in which the business is organized. Some of the desirable characteristics of the small organization can be retained within a large company if the business is organized into blocks, products, or families of products or income-producing services, or of staff functions.
3. Innovation. Among some 400 highly motivated managers, 73% said that their ideas were usually adopted. Of approximately 400 poorly motivated managers, 69% said their ideas were not used. Part of the reason the upper group is highly motivated is that its members exist in a climate that is geared for change.
4. Fluid communication. As organizations grow, communication is usually formalized. This formalization can actually result in communication breakdown. In a small shop, June yells something across the room to Harry; information is exchanged, and that is all that is needed. But in a growing company, memos are dictated that are typed in duplicate, filling

up cabinets and taking up floor space. People are hired to deliver the memos to distribution points, from which they go into baskets, tying managers who must read and answer these communications to their desks. The growing mountain of paperwork worsens until some agile competitor comes along who does not have this paper problem, and who threatens the growing company by giving superior service at a lower cost. (This problem may even be exacerbated by computer communications.)
5. Goal orientation. This differs from authority orientation. When a person is asked why he or she is doing something and the reply is, "Because the boss told me to," that is authority orientation. But if, in reply, the worker explains the job and what it will achieve for the organization, that is goal orientation.
6. Stability. This means many things, one of which is job security. The worker must feel assured that there will be a job for him or her. This kind of stability is essential.

Rensis Likert, one of the most famous of the behavioral scientists, also discussed climate to a large degree when he described his *system 4* kind of company. He isolated three variables that are representative of his total concept of *participative management:* (1) the use of supportive relationships by the manager, (2) the use of group decision making and group methods of supervision, and (3) the manager's performance goals.

The use of group decision making and supervision does not necessarily mean that the group makes all decisions. Rather, the emphasis is on the involvement of people in the decision-making process to the extent that their perceptions of problems are sought, their ideas about alternative solutions are cultivated, and their thoughts about implementing decisions that have already been made are solicited. Likert's participative-management process can occur on either an individual or a group basis. Participative management involves the integration of people around production. It does not mean that people take a back seat to production at any cost. An organizational climate in which psychological satisfaction and feelings of achievement are probable facilitates accomplishment on the job.

Likert measured the relation of supportive and participative management to productivity. He found strong evidence suggesting that the organization that exhibits a high number of supportive relationships and that uses the principles of group decision making and supervision (where there are high performance aspirations) has significantly higher levels of achievement.

Likert made a distinction between what he called *casual, intervening,* and *end-results* variables. The casual variables refer to different management-systems characteristics as follows:

System 1—explorative authoritative
System 2—benevolent authoritative

System 3—consultative
System 4—participative group

Likert explained each variable in detail in *The Human Organization*. The intervening variables—such as loyalty; performance goals of subordinates; degree of conflict versus cooperation; willingness to assist and help peers; feelings of pressure; attitude toward the company, job, and superior; and level of motivation—are of key importance. The end-results variables refer to tangible items such as volume of sales and production, lower costs, higher quality, and so on (i.e., results).

Participation is important in gaining employee commitment on an overall basis. It can lead to less need for the use of formal authority, power, discipline, threat, and pressure as means of eliciting job performance. Thus participation and its resultant commitment become a positive substitute for pure authority. Commitment may be much harder to achieve initially, but in the long run it may prove much more effective to gain it this way.

As mentioned previously, Likert's description of participation is really a description of climate, an assessment of which is fundamental to safety-program success. Climate contributes to employee attitudes and to acceptance of the safety program. It also contributes to credibility, which is necessary for effective communications with employees about safety. Although hard to define, climate is important.

What builds climate? Myers suggested six components and Likert ten. Scanlon spelled out six different ones. Each of us might be able to identify others from our experience. Anyone who has worked in an organization knows that climate exists and knows its importance. There are companies with vastly differing climates: from permissive to tightly controlled, from free to guarded, from creative to stifling, from rigidly scheduled to flexible, and, finally, from safety-minded to lax. In addition, different safety programs generate a variety of climates. The organizational climates described are real. They are a product of both the corporate climate and the functioning safety program. Scanlon, Myers, and Likert, in their discussions of healthy climates, invariably mentioned two key factors: participation and organizational goals.

Analyzing Climate

Both the climate of a safety program and corporate climate are difficult to assess—easy to feel, but often difficult to get a handle on. To assess safety program climate, it may be best to question other people in the organization, notably line managers and workers. One way to do this is with a questionnaire. Asking employees for their ideas and input not only furnishes a good indication of safety-program climate, but also elicits some excellent ideas for the improvement of the program. Corporate climate can also be gauged by a questionnaire. There are devices available for this type of assessment.

Climate seems to be highly important to any safety program. A safety program is most effective in organizations where there is credibility between management and worker, where there is mutual trust and respect, where there is an honest working relationship among managers at all levels, and where a basis has been established for the agreement reaching and objective setting that are vital to a safety program. The closer an organization is to Likert's system 4, the easier it is for people at all levels of the organization to get safety results. The more closely a company is aligned with the system 1 organization (the authoritarian, hierarchical, bureaucratic approach), the more difficult it is to achieve a working safety program.

The Boss's Style

Other key influences on employee attitudes and behavior (actions) are whether the job is important and meaningful to the employee, the degree of employee involvement and participation in the important aspects of the job, and how much recognition is given for good performance. These insights come from the thinking of Frederick Herzberg, Chris Argyris, Rensis Likert, and other influential management theorists.

Conflict Theory

Argyris's theory provides safety managers with insights into why people commit errors. Argyris took human nature as a starting point and analyzed the process of growing up and maturing. Children, he said, are passive and depend on their parents; they exhibit few behaviors. Their interests are shallow and short-term. They are at all times in a subordinate position in their relationships with older persons, and they are relatively lacking in self-awareness. As children mature, this changes. Adults are active, independent creatures who like to stand on their own. They exhibit many behaviors, and their interests are deep and long-term in nature. Mature people view themselves as equals in most relationships, not as subordinates, and they have self-awareness.

According to Argyris, all organizations—whether industrial, governmental, mercantile, religious, or educational—are structured under certain principles:

1. They have a chain of command. This creates a superior–subordinate relationship that, in turn, causes workers to be dependent on the boss, to become passive, and to lose interest in the job.
2. The span of control is small. This creates dependency and reduces the freedom and independence of the worker.

3. There is a unity of command, or only one boss. This creates dependence and highlights the subordinate role of the worker.
4. They are characterized by specialization. The work is broken down into small, simple tasks. This leads to a lack of interest, to a lack of self-fulfillment and feelings of self-importance, and to dependence and passivity.

These four principles create *managerial pyramiding,* with a boss, a series of minor bosses under that boss, a series of minor-minor bosses under them, and finally the workers, who would like to consider themselves equals but find that they are very unequal. Such structures make workers highly dependent on their bosses.

Argyris says that these principles of management are, in fact, in conflict with the needs of individuals, and he suggests that this conflict causes people to quit, to become apathetic about their jobs, to lack motivation, to lose interest in the company and its goals, to form informal groups, to cling to the group norms instead of the company's established norms, and to evolve a psychological *set,* a belief that the company is wrong in most things it attempts to do. It also causes accidents that result from inattention, disregard of safety rules, poor attitudes toward the company and safety programs, and so on. The normal management reaction to these symptoms is to establish more control, more specialization, and more pressure.

What can we do about this? Since we cannot feasibly change mature people into immature ones (nor would we want to), the only option is to look at the organization and see how we can change it. This leads us to organizations (and safety programs) with less control, less pressure, and fewer superior–subordinate relationships.

Argyris, one of the most famous of the behavioral scientists, proposed *leveling,* or the use of group decision making and supervision so that the boss does not necessarily make all decisions alone. The emphasis in leveling is on the involvement of people in the decision-making process to the extent that their perceptions of problems are sought, their ideas on alternative solutions are considered, and their thoughts on implementing decisions that have already been made are solicited.

Likert's Theory

Another theory is that of Likert, whose studies concerned the effect of the supervisor–employee relationship on productivity. Among his findings were the following:

The tighter the supervisor's control over the employee, the lower the productivity.

The more the supervisor watches and supervises the worker, the lower the productivity.

The more punitive the supervisor is when the employee makes a mistake, the lower the productivity.

In short, Likert's research indicated that if managers want productivity, they should not "control" employees. Controlling employees will cause them to work less. Today, management accepts these behavioral theories, and certainly they can help safety directors increase the acceptability and productivity of a safety program.

The relation of employee motivation to measurements of employee productivity is not strong; in fact, there could even be some negative correlation between motivation and measurement. The more managers tend to measure and control employees, the less work they are likely to get from them. Employee motivation is better achieved by means of peer pressure, by treatment of employees as mature and intelligent human beings, and by less control.

We have not yet been very successful in applying these theories to safety management. We have not really "turned employees on" to our safety programs. We may wish to evaluate a safety program by asking employees what it means to them. Often a safety director who does this finds that the employees think the safety program consists of the list of work or safety rules on the bulletin board, or of some silly posters or boring safety talks that might be endured periodically. These are typical employee reactions to a traditional safety program.

Job Motivational Factors

A major researcher in the area of job motivation, Frederick Herzberg developed the *motivation–hygiene* theory. Herzberg called certain variables *hygiene factors* and others *motivation factors*. By improving the hygiene factors (company policy, supervision, interpersonal relations, status, etc.), a company can make a dissatisfied worker into a satisfied one, although this does not necessarily mean that the worker will be motivated. The motivation factors (achievement, recognition, the work itself, responsibility, etc.) have to do with the job itself, while the hygiene factors are peripheral to the job.

The following things determine the worker's level of satisfaction:

1. Money
2. Status
3. Relationships with the boss
4. Company policies
5. Work rules
6. Working conditions

The following factors determine the worker's level of motivation:

1. Sense of achievement
2. Recognition

3. Enjoyment of the job
4. Possibility of promotion
5. Responsibility
6. Chance for growth

If we structure the safety program only the basis of the worker's level of satisfaction, people will never be excited by it. They probably will be bored by it. On the other hand, if we build the safety program around the worker's level of motivation, it has a good chance for success.

Individual Differences—Personality

Differences exist not only among individuals but also among groups. There usually is a definite difference, for instance, between the personal values and attitudes of the younger worker and those of the older worker. This is important to safety professionals, who must sell their safety program to both older and younger workers.

Why are younger and older workers different? They are different because their values and attitudes were determined by different institutions and different life experiences. Consider the changes that have occurred in recent years in the three basic institutions of the family, the church, and the school:

1. Family influences have changed considerably. More mothers work outside the home, more parents travel in their work, and more families have become fragmented as family members move far from one another.
2. Religious institutions have less influence than they once had and appear to be losing ground steadily.
3. Schools have changed their teaching methods, and they, too, appear to be losing influence. Old teaching methods based on memorization have given way to a variety of educational concepts.

Furthermore, the war in Vietnam and more recent conflicts, the nuclear threat, the civil rights and human rights movements, concern over the environment, and especially the communications explosion have all contributed to changes in values and in attitudes. The young today have a distaste for social and institutional rigidity. They fear the depersonalizing effects of technology. They are intolerant of hypocrisy. They have different life styles from earlier generations and a different work ethic. The credo of the older generation was that we live in order to work. Many of the younger generation believe that we work in order to live. Younger workers, for the most part, do not like doing things the hard way if there is a simpler way. Often they resent it when the boss says, "Do it my way."

They often balk at safety rules. They believe in being able to "do their own thing."

Although younger people rely on institutions to do the things they cannot do themselves, they seem to have lost confidence in the large institutions of government and business. Where does this leave us? It leaves us with young people who have different values and attitudes from their elders. This discrepancy is not going to go away. Managers have to accept it and live with it. Managers also have to reconsider their styles of leadership and their training approaches. Future managers will be using more employee-centered leadership approaches and more participative styles of management. Safety training in the future will have more to do with understanding and dealing with people than with teaching technical subjects. Thus, it is common in safety training today for supervisors to teach means of understanding people better.

Peer Pressure

At the employee level, the primary motivating influence is peer pressure, or pressure from the informal group. Although each employee is an individual, he or she is also an integral part or member of a group, and every manager must take this into account.

In chemistry, elements combine to form substances that have entirely different properties from those of the individual elements. People combine to produce groups that have entirely different properties from the individuals in them and from other groups. Each group has a distinct personality of its own.

Each group makes its own decisions. It sets its own work goals, which may be identical with or different from management's goals. Each group sets its own ethical standards. For instance, a group might decide that stealing things from the company is allowable and hence would not exert pressure on individual members who take small items (pencils, notebooks, hammers, etc.). A group member caught stealing from another group member, however, is usually in deep trouble. The group takes care of the punishment itself. Management does not have to because the group's ethical standards state that is wrong to steal from another group member. The group has decided, and the group enforces its decision.

The group also sets its own safety standards, which it lives by, regardless of what management's standards are. Take, for example, hard hats. If a group of construction workers makes the decision that hard hats should be worn, all members will wear hard hats. If, however, the group decides against head protection (which groups did some years ago), it will exert pressure on members not to wear hard hats.

How can we cope with the phenomenon of the group? First, we try to understand the groups that we have in our company; second, we try to identify the respective strengths of those groups; and third, we try to build strong groups with safety goals that are the same as ours.

The Union

The union may or may not influence employee motivation, depending on whether or not there is one, how strong it is, to what extent it represents the feelings of the workers, and so on. The union also influences workers to work either safely or unsafely, a subject we will discuss further in the next chapter.

ATTITUDINAL STATE

In his book *Supervisor's Guide to Human Relations,* Earle Hannaford defines attitude as the potential for action, and safety attitude as a "readiness to respond effectively and safely, particularly in tension-producing situations." He goes on to state the three components of attitude:

$$\text{attitude} = \text{learned responses} + \text{habit} + \text{emotional set}$$

and suggests there are four steps in building attitudes (Table 11-1).

Hannaford's work with attitudes has shown the relationship between attitudes and results in the area of safety (Figure 11- 2). As is readily evident in the graph on the left, the poorer the safety attitude of the employee, the greater the number of lost-time accidents that occurred during the five-year period studied. The 769 male employees Hannaford studied came from 47 companies representing a cross section of various industries—companies with excellent, average, and poor safety records. The graph on the right shows that, as the supervisor's safety attitude test score worsens, the number of lost-time accidents per employee increases. Obviously, the main conclusion to be drawn from this study is that a positive attitude toward safety fosters safe working practices.

AROUSAL STATE

In Figure 10-8, the effect of arousal was depicted. When the task difficulty is held constant, the level of performance increases with increased arousal. Since, in the context of human-error reduction, increased performance means fewer errors, increased arousal means fewer errors. In the context of overload, the

Table 11.1. Practical safety-program activities that build good attitudes

Four Steps in Attitude Formation	Typical Safety Activities to Use
Step I Laying the Foundation for the Attitude	**Mass Media** Safety slogans, safety posters, safety talks. Motion pictures and sound strip films of general safety nature. Training classes and demonstrations for groups on job methods and theory. Company safety policies. Safety contests and competitions of a group or company nature.
Step II Personalizing the Attitude for the individual	**Learned Responses and Habit Formation** On-the-job training in correct safe work methods. Good supervision—immediate correction of violations of safe working practices to build safe habits. Individual participation in safety meetings, safety planning and safety inspections. Motion picture and sound strip films dealing with job methods and sequences. Recognition of personal contributions by boss and higher authority figures. Individual safety awards
Step III Fixation of the Attitude Emotional Set	**Emotional Set** Discussion of actual job-related accidents with individual participation. Role playing—permits identification with the projection of self by individual. Motion pictures with high emotional content relating to safety in general and to job performance. Actual demonstration of their personal interest in safety by the boss and higher management—making it the No. 1 item—catching the attitude from authority figures.
Step IV Keeping the Attitude Alive	**Attention, Memory and Emotional Set** Checkup on attitude status of individuals and groups using industrial safety attitude scales for employees and supervisors to see where emphasis is needed. Attitude surveys.

Note: Plan safety program to offset "Safety program fatigue" by using some of the items designed to provide for Steps I, II, and III since employees may be in any one of these steps or may have regressed from III to II, or II to I.

Source: E. Hannaford, *Supervisor's Guide to Human Relations*. Chicago: National Safety Council, 1967. Reprinted from *National Safety News,* a National Safety Council publication, 444 N. Michigan Ave., Chicago, IL.

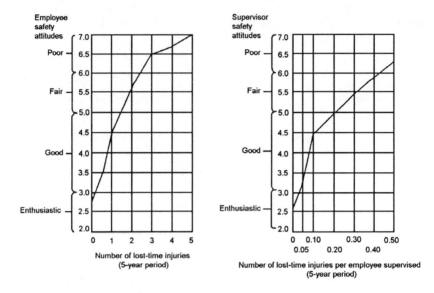

Figure 11-2. Employee and supervisor safety attitudes. (From E. Hannaford, *Supervisor's Guide to Human Relations*. Chicago: National Safety Council, 1967. Reprinted from *National Safety News,* a National Safety Council publication, 444 N. Michigan Ave., Chicago, IL.)

effect of increased arousal is to enable a person to take a greater degree of overload. Human beings can be overloaded more and longer if aroused. While this is a fact that we ought to be aware of, it does not seem to lead us to any particular control device. We cannot, for instance, build mechanisms to keep workers constantly aroused in order to make them permanently better able to take greater overload.

BIORHYTHMIC STATE

Biorhythms are the 23-day physical, the 28-day sensitivity, and the 33-day intellectual sinusoidal cycles that, theoretically, originate on the day a person is born and cycle uninterrupted until that person dies. Accident-prone days or periods are said to exist for approximately 24 hours when any one of the three sinusoidal curves passes through 0 degrees or 180 degrees of the sinusoidal cycles. These accident-prone periods represent approximately 20% of a person's life. Biorhythms are not intended to predict when an accident will occur; they only indicate a period of time when a person is more prone to causing an accident.

Whether or not we believe in this concept, and there are conflicting research findings, it must be included in our thinking about the state of an individual.

Perhaps the real question is whether or not biorhythms do affect accident proneness. There is a great deal of information currently being promulgated by the proponents of biorhythms. To date, most of the studies made to test the concept have been performed using methodologies that lead to no real conclusions. There is no doubt that the cycles exist, but we simply are not sure whether they are related to accident occurrence. Typical of the studies that support the biorhythm concept is the study made by Jacob Sanheon, at the Naval Weapons Support Center Quality Assurance Department in Crane, Indiana. He evaluated 1,308 accidents from three separate sources: the industrial accidents at the center, the community accidents from the local emergency room, and the motor vehicle fatalities (single-car crashes) in the State of Indiana for a seven-month period, and found that 41.75% of the accidents occurred on accident-prone days and 52.03% occurred during accident-prone periods (constituting 20% of the time).

Typical of findings against use of the biorhythm concept is an investigation of biorhythms made by the Workmen's Compensation Board of British Columbia in 1970. The board found no correlation between critical days and accident occurrence.

WHAT TO DO ABOUT STATE

It appears that there is a considerable amount of activity that management can pursue to influence the state aspect of overload. It can, first of all, engage in a myriad of activities aimed at improving the motivational field of the employee, including enriching jobs, allowing employees to be more involved in the decision-making process, and allowing employee participation. Management can also engage in a process of climate building. Many of these activities will be discussed in the next chapter.

REFERENCES

Argyris, C. *Personality and Organization.* New York: Harper & Row, 1957.
Geller, E. S. The psychology of occupational safety, in *Industrial Safety and Hygiene News,* Jan. 1992.
Hannaford, E. *Supervisor's Guide to Human Relations.* Chicago: National Safety Council, 1967.

Herzberg, F. *Work and the Nature of Man.* Cleveland: World, 1966.
Likert, R. *New Patterns of Management.* New York: McGraw-Hill, 1961.
Likert, R. *The Human Organization.* New York: McGraw-Hill, 1967.
Myers. M. The manager's role in motivation. Paper presented at Western Printing and Lithographing, Racine, WI, 1966. Photocopied.
Pattenden, T. Using Management and Motivation in Safety. Paper presented at OSH 1994, Toronto, Canada, October 1994.
Petersen, D. The safety profession: concepts and programs, *Professional Safety,* December 1984.
Sanheon, J. Biorhythm analysis as applicable to safety. Paper presented at the National Safety Congress, Chicago, September 1975.
Scanlon, B. Philosophy and climate in organizations. *ASTME Vectors,* Vol. 5, 1969.
Social Research, Inc. *The Effectiveness of Safety Education Materials.* Chicago: Social Research, Inc., 1962.
Workmen's Compensation Board of British Columbia. *An Investigation of the Biorhythm Method.* Vancouver: The Board, 1970.

Chapter *12*

Reducing Overload-Caused Errors

Chapter 8 discussed the major causes of error in the categories of psychosocial and personality causes. Some of these factors are rooted in the organization, such as frustration and overload.

Chapter 8 also concentrated on job stressors:

- Job satisfaction
- Physical conditions
- Organizational factors
- Work load
- Work hours
- Work role
- Work task
- Career development

In addition, Chapter 8 covered some specific stressors of the modern industrial scene:

- Takeovers and mergers
- Downsizing
- Lack of loyalty to the employee
- Lack of control
- Lack of social support
- Incentive programs

There are clearly many job stressors, and changing all of them would be a mind-boggling task. Some we have little or no control over (takeovers, mergers, downsizing, lack of loyalty). Others we control to some extent.

179

This chapter discusses some of the many options and approaches you might use to "fix" the organization.

Chapter 8 emphasizes psychological overload, but overload can also be physiological or physical. Physiological overload falls mostly under the purview of the industrial hygienist. There are a plethora of standards available to us to tell when a person is physiologically overloaded. Physical overload comes under the purview of the ergonomist or the industrial engineer. Here again there are a myriad of standards to assist us in determining how much is too much for the average worker. Therefore, this chapter will concentrate mostly on psychological overload, which today is under no one's purview, and is the largest problem in human error.

The management of overload involves reducing the stressors, changing the person, coping with the stress reaction, or becoming ill.

Reducing overload at its source might include these kinds of strategies:

- Conflict reduction (identification, confrontation and resolution)
- Decision making
- Assertion of self
- Opening of communication
- De-fuzzying
- Meditation
- Time management
- Reducing work loads
- Reducing time pressures
- Delegating
- Changing work activities
- Changing positions
- Taking time off
- Redesigning jobs
- Others

Helping people to cope with overload might include these approaches:

- Moving from type A to type B behavior
- Psychotherapy
- Assessing strengths and weaknesses
- Acceptance of self
- Improving sense of humor
- Developing better coping skills
- Physical exercise
- Weight control
- Progressive reaction techniques (PRT)
- Vacations (removal from the situation)
- Nutrition

Reducing Overload-Caused Errors

- Escapist activities
- Creating (painting, for instance)
- Hobbies
- Venting aggressions
- Hypnosis
- Play

These approaches should be avoided:

- Reliance on alcohol, drugs, food, smoking, or caffeine
- Denial, ignoring the problem

The choice of the strategy is ours. We can choose to deal with overload at the source, to change ourselves, to cope better, or to be consumed by it.

The above are some general strategies. Some specific ones follow.

STRATEGIES FOR CHANGING THE WORK ENVIRONMENT

At the corporate level the issue is how can overload be reduced voluntarily within the company. There are three approaches:

1. What the company is willing to do.
2. What the union will do.
3. What workers themselves will do.

There is a lot the individual can do, and many of the general strategies already discussed are individual strategies. Thus far it seems that unions have done little—they apparently are either disinterested or oblivious to the problem. The company can do much if it chooses to. It can reduce job-caused stress, and it can assist the individual workers with programs that will help them cope with the overload.

There are four basic paths to follow in treating stress symptoms:

1. To eliminate the stressors or remove the person from the overload.
2. Pharmacotherapy—treatment with drugs.
3. Psychotherapy—treatment with counseling.
4. Exercises and physical training to help workers cope.

STRATEGIES FOR REDUCING OVERLOAD

The following are a number of specific remedial interventions:

- *Job redesign,* modifying the content of work, enriching the tasks done, or rotation.
- *Organizational modification,* giving greater autonomy, more ownership, more delegation of decision making and problem solving to the worker.
- *Ergonomic redesign,* using what we know to make the job user-friendly.
- *Modifying the working space and time,* removing crowding or isolation, allowing rest periods.
- *Reducing forced overtime* through better planning of personel usage and better scheduling.
- *Providing more information* on everything, so the workers can feel they are a part of the organization.
- *Allowing worker input* before changes are made in the work set-up.
- *Allowing participation* in most decision making.
- *Providing training.*
- *Providing psychological first aid.*

The following strategies also will reduce the stress the organization places on individuals:

- Goal setting
- Autonomous work groups
- Performance feedback systems
- Role specification
- Training programs
- Climate/culture improvement
- Conflict resolution strategies
- Employee surveys
- Wellness programs

In sum, the above-listed organizational strategies have to do with improvement of the quality of work life through goal setting, participation, job enrichment, improving the culture and climate, performance feedback, specification of roles as well as responsibilities and accountabilities (defuzzying things), surveying employee perceptions, and use of training programs, wellness programs, and employee assistance programs.

We will touch on a few of these strategies and some others. First are a number under our direct control that have to do with fixing our management system.

Goal Setting

When employees are involved in goal setting, a number of good things happen:

- They are a part of the process; they have a say in what will be going on.
- Thus, they feel they have some control—perhaps the single most important ingredient of stress reduction.
- This ensures results, as the goal is theirs, not just management's.

Work Scheduling

Work scheduling is a severe problem in many organizations. Here are a few examples:

- A large computer manufacturer doubled production in one year because of sales, and did so with only a 10% increase in head count. They did it with overtime, and to a large degree, forced overtime. The result was that the accident frequency rate doubled. Since this was the early 1980s (largely the pre-stress era), they had a large increase in subjective injuries and an increase in objective injuries caused by physical, physiological, and psychological fatigue. In one department there was a rash of dermatitis claims. The department had more forced overtime than most, and employees requested transfers out—all of which were declined. The employees' answer was to stick a hand in a solvent, get a rash, and get off—and then, to be transferred.
- A large food processor went into the busy season with a management edict to maintain a lower headcount. Forced overtime resulted in a total deterioration of an excellent safety record. Accident costs skyrocketed (management was not measuring that), but headcount remained the same (management was measuring that). Stress claims emerged along with all other injuries.
- Another company was coping with downsizing (who isn't?), which meant more work for fewer people and forced overtime—as much as 12-hour shifts, seven days a week. One employee had 16-hour shifts and had to drive 42 miles each way each day. One night he fell asleep at the wheel, hit a bridge abuttment, and was killed. The result was a multi-million dollar liability suit against the company. They lost.

Work scheduling is a particular problem in some industries. Where production is scheduled by customer demand, the company has less control. In these cases the workers have even less control; at times they leave one shift at midnight only to be called in at 4:00 A.M., which not only is physically hard but is disruptive to family life also.

Joe Ladou, of the University of San Francisco, has done major research in this area and offers these criteria:

- Single night shifts are preferred over consecutive, repetitive night shifts, because the shorter exposure does not alter circadian rhythms significantly.
- Each period on a night shift should be followed by at least 24 hours of free time to offset the harmful effects of sleep deprivation. This period should also be applied to early morning workers if their sleep routines are disturbed by their shift hours.
- The length of a shift should depend on the type of work to be performed. Shifts involving light or uncomplicated tasks can be longer than those requiring heavy physical work or difficult mental efforts.
- Older workers should not be placed on longer shifts (10 or 12 hours) because of the added stress they experience.
- The length of the rotating shift cycle should not be too long—for example, 4 weeks is better than 40 weeks on the same shift.
- Shifts should be rotated regularly so that workers can plan their family and social lives.
- When rotating shifts, it is preferable to rotate workers forward to a later shift than to an earlier one.
- Workers on permanent night shifts should be given as many free weekends as possible so that they can participate in some family and social functions.

Performance Feedback Systems

How would you like to go bowling where no score is kept, where a large curtain hangs between you and the pins, and where you hear no sound of the ball hitting the pins? Probably you would quit bowling. What if you could not quit? You might get discouraged and frustrated, and begin to head for an illness. This is what lack of a performance feedback system can do.

This happens constantly in industry. Workers receive no feedback, except perhaps at year end (and sometimes not even then). In a recent survey to find the strengths and weaknesses of the safety systems of four large railroads, lack of recognition was found to be the single most important element of the safety program by a wide margin. Our inability to provide feedback is a great failing in American management, and it leads to frustration, loss of motivation, reduction of productivity, more accidents—and stress-related illness.

Performance feedback starts with role definition.

Role Definition

Modern management theory puts great emphasis on removing role ambiguity. The more clearly we can define individual roles (what people are supposed to do), the more likely it is they will perform them. The fuzzier we keep things, the less likely they are to perform, and the more we keep them in a stress situation.

Most newer management techniques start with role definition, whether it be management by objective (MBO) or one-minute managing.

Training

Teaching people how to do what is expected of them also helps. Knowing how they can do it increases their comfort level. Training does not beget performance, but it does allow it. It also helps to alleviate stress by putting control into the hands of the workers—who now know how to do the tasks, and that doing them is their responsibility.

Discipline

Discipline can be a major contributor to a stressful environment, particularly when it is not consistent. When workers are zapped periodically for behaviors that are condoned at other times, or when performed by other people, a stressful environment is created.

The more that antecedents are laid out (you will be zapped if you do this) and the more that consequences are consistent (he does it and is zapped), the less the behavior will occur, for you have given the person the button that stops the pain.

Fixing the Climate

Another broad approach is fixing the climate. Here are some general thoughts on climate and several specific ideas that usually do a lot to build a healthy climate, providing participation and job enrichment.

Climate is a major influence on the behavior of both managers and employees, but a basic problem is that the employees' perception of the organization's climate and philosophy may be different from what management intends. Two possible factors may contribute to this difference in desired versus actual perception. First, perhaps not enough effort has been expended in communicating the guiding philosophy down the line. The second factor may be a discrepancy between what is professed and what actually occurs. The individual's closest point of contact with the organization is his or her immediate superior. If the superior's actions do not reflect the organizational philosophy, a perception discrepancy occurs.

Some of the basic climate requirements for maximum individual performance in an organization are:
- There must be central overall goals or objectives toward which the organization is striving.
- The objectives must be communicated down the line with the idea of getting commitment to value, reasonableness, and feasibility.
- Functional areas, departmental units, and individuals must also have specific goals to attain. These must be derived from the central goals, and interrelationships must be perceived.
- The interdependency of all subunits within the organization in the accomplishment of results must be clearly established, and a framework for interunit cooperation must exist.
- Meaningful participation on the part of the individuals should be the key. This means participation in the sense that the individuals have a "real" part in determining job objectives.
- There must be freedom to work in the sense that the workers have an opportunity to control and adjust their own performance without first being exposed to authority and pressure from their superiors. In addition, they must receive support and coaching from their superiors.

Participation

Clearly, one of the biggest aspects of climate is the amount of participation that is allowed. Participation casts each worker into a role quite different from his or her role in a hierarchical program. The change to a participative program changes workers from inferior subordinates to equals, peers. It changes them from persons who receive orders and do the work as ordered to ones who help decide what is to be done and who then act in accordance with their own wishes, decisions, and goals. The workers, under participation, live different roles, and thus mold new values, new attitudes, and changes in behavior. Under participation, the workers want to achieve those things that they have decided need to be done. They, in short, become motivated individuals because their roles are the roles of persons who want to accomplish specific results. New roles also place them in a position with their peers of having to and wanting to cooperate. The chosen goals are group goals, not goals thrust on them and their superiors. Since the group has set the goals, achieving those goals is accepted; it is the group norm. Behavior that leads to group goal achievement is normal behavior. Behavior that does not lead to this becomes a deviant behavior.

Figure 12-1 summarizes the findings on participation. People who report high levels of participation are less likely to experience mental strain in the form of job dissatisfaction, a job-related threat, or low self-esteem. The lack of role ambiguity that accompanies participation apparently enables the workers to better utilize their

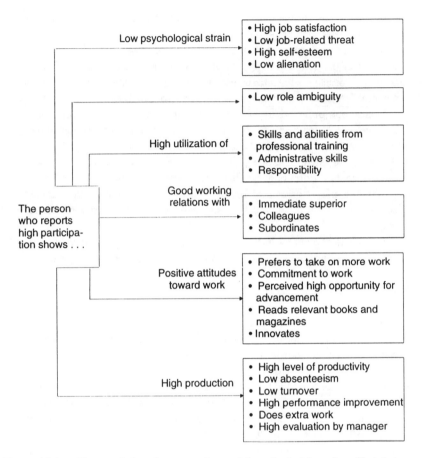

Figure 12-1. Characteristics of persons who participate in decisions that affect their work.

skills and abilities in performing their work. High participation also is accompanied by high responsibility.

Phyllis M. King summarizes the role of control, participation, training, and empowerment:[1]

> With the rise in office and service occupations, and a parallel decline in manufacturing, concerns about work organization and stress have become serious occupational health issues. One indication of this trend: Stress-related

---

1. Reprinted with permission from P. King, The psychosocial work environment: implications for workplace safety and health, *Professional Safety,* March 1995.

disability claims are the most rapidly growing form of occupational illness within the workers' compensation system.

Studies of work settings have described the deleterious effects of deskilling, loss of freedom, passivity, social isolation and related aspects of alienation. Research has shown that these factors negatively impact psychological and physiological functioning, as well as nonworking life.

It has been observed that employees with little control over their jobs become passive in other aspects of their lives; alienation at work contributes to larger political and social alienation. In contrast, employees with a greater degree of self-determination and control over work demonstrate greater involvement in addressing and changing problems encountered.

Working life is of central importance to many people. If they cannot control significant aspects of the work and work situation, risk of learned helplessness arises.

Jobs characterized by little control, influence, learning and development entail risks of learned helplessness and, consequently, depressive reactions.

DIMENSIONS OF CONTROL

Individual and collective control. When assessing the social character of a work process, the issue of individual autonomy within a collective process is central (although studies linking the social character of work to stress are limited). Two studies found some indications of the relationship between democratization (individual and collective control) and stress. In autonomous groups, production demands produced increased pressure, while nonautonomous groups in similar situations experienced higher stress levels both on and off the job. Fatigue after work was considerably more common among those in nonautonomous groups as well. Therefore, according to Gardell and Svensson's interpretation of this contradiction, autonomy is a means of coping with increased workload.

Vertical and horizontal control. Distinction between control within and control over a situation corresponds to distinction between horizontal and vertical job discretion in organizational theories. Horizontal/vertical work dimensions can be used when initiating or analyzing different types of work reform activities.

Job rotation and job enlargement are limited to the horizontal dimension and reflect control within. Job change that combines qualitatively different tasks (i.e., task planning and execution) represents reform that influences the vertical dimension. Several studies have found that employee effectiveness increased when an organization's management strengthened vertical control, while simultaneously surrendering some horizontal control to the employee or employee collective.

DIMENSIONS OF PARTICIPATION

Informal vs. formal. Cotton, Vollrath, et al found that informal participation had the greatest correlation with both increased performance and job satisfaction. Participation was nurtured by strong interpersonal relationships between employees and management.

Direct vs. indirect. Most research has focused on representative participation, an indirect form of participation. In such systems, central management retained almost complete control. Although productivity did not improve, satisfaction, at least among representatives, did increase. Conversely, direct participation correlated with increased productivity and affective response.

Long term vs. short term. Studies have found that organizations with short-term individual involvement had less commitment than those where participation in decision making was long-term (weeks or months). Lawler emphasized the need for a long-term approach because short-term outcomes are often either nonexistent or negative.

THE ROLE OF ADULT EDUCATION

The traditional safety and health approach has been to transform work environment problems into narrow concerns, with technical issues of measurement determined within the context of the legal enforcement system. Workplace safety and health has become an issue for experts and inspectors, yet employees often play a passive role. This turns them into "objects" whose welfare is the responsibility of others.

Adult education adjusts this focus, transforming employees into "subjects" whose actions can shape their fate. This view maintains that the purpose of education is to promote participation of people who, as a result, gain control over their personal and community lives. This requires treating employees as creators of their own learning. Via dialogue about real-life issues, employees gain critical understanding of causes of workplace problems (socioeconomic, political, cultural and historical forces) and their role in challenging those forces. "Development of critical thinking" and "education for empowerment" are the focus. These concepts surpass mere understanding of health problems; they call for employee participation and action as ways to improve working conditions and, thus, safety and health.

WORKER EMPOWERMENT

As an approach to learning, education for empowerment is participatory and based on real-life experiences; incorporates dialogue among educators and workers; critically analyzes organizational and system-wide causes for problems; and facilitates employee action and empowerment. These principles follow basic tenets of adult education theory: 1) People retain information best when actively involved in problem-solving exercises and hands-on learning. 2) Education is most effective when it includes the context of behaviors, including analysis of obstacles to safe work practices.

Use of participatory methods in training is a first step toward empowerment. Participatory methods should help employees develop critical thinking, practice social action skills and gain confidence needed to actively improve the work environment.

APPLYING EMPOWERMENT EDUCATION

Worker safety and health education is now recognized as a key component of preventive occupational health programs. Some 100 OSHA standards mandate training, as do requirements from regulatory bodies such as Joint Commission on Accreditation of Health Care Organizations. However, no standards mandate quality or excellence in training; no criteria has been established for appropriate teaching methods, organizational structures or program evaluation.

Maturity

Another very important climate factor is whether or not the organization treats its people as mature human beings. Most adults are motivated to be responsible, self-reliant, and independent. These motives are acquired during childhood from the educational system, the family, and communications media such as books, television, and radio. But the typical organization confines most of its employees to roles that provide little opportunity for this, because of typical management beliefs that workers need extensive supervision.

Conflict Resolution

How an organization deals with conflict also is an important factor. Figure 8-1 described some of the reaction options to conflict.

It is obvious that conflicts are a part of workplace problems. Figure 12-2 shows several conflict resolution strategies.

In the last chapter a number of individual stress reduction strategies and how the organization can assist individuals in them were discussed. In this chapter some organizational strategies were discussed. Let us summarize this chapter with the checklist in Figure 12-3. We urge you to assess your company's progress.

```
                           AVOID
• Ignore
• Keep from happening

                          DEFUSE
• Cool down for rational discussion
• Resolve minor points/avoid major issues
• Keep issues clear
• Look for conflict of intent, misunderstanding

                         RECONCILE
1. Diagnose
     a. Your own intent
     b. Type of conflict
     c. Relation between members
2. Initiation
     a. When and how to approach, confront
3. Listening
     a. Uncover specific needs, concerns of each
     b. State what's important to you
     c. Check what's important to other person
4. Problem solving
     a. Invite/suggest alternatives
     b. Temporarily alter restrictions
        —Remove restrictions
        —Impose restrictions
     c. Define problem
     d. Brainstorm
5. End discussion
     a. State your understanding of what has been said and why (if resolved); or
     b. Make decision, acknowledge right to differ, explain what decided and why (if
        resolved); or
     c. Suggest you continue discussion another time (if emotions are intense)
```

Figure 12-2. Conflict resolution strategies.

```
COMPANY _____    DATE _____

AREAS                                PROBLEMS/CONSIDERATIONS

 1. Have you had a climate/stress survey of any type?
 2. What problems were identified?
 3. Are you/have you been in a merger or takeover?
 4. Are you downsizing?
 5. Are you losing your employees' feelings of loyalty?
 6. Do you have an ongoing job enrichment program in place?
 7. Where (at what level) are most decisions made?
 8. Do you have an ergonomic analysis program in effect?
 9. Do your people work overtime? Is it forced on them?
10. Is there an effective effort to employee participation?
11. Do you use MBO?
12. Are you experimenting with autonomous work group?
13. Do you have an active wellness group?
14. How did you score on the Likert Climate Scale?
15. Are your people treated as mature individuals?
16. How does your organization deal with conflict?
```

Figure 12-3. Stress assessment.

Having answered these questions, now randomly select 25 employees and ask them the same questions. Once you have positive responses from everyone, relax. You probably will not have too many stress claims this year (unless you introduce change, start downsizing, or are in a large city with too many lawyers and doctors).

REFERENCES

Argyris, C. *Personality and Organization.* New York: Hayron, 1951.

Cotton, J. L., D. A. Vollrath, K. L. Froggatt, M. L. Hengnick-Hall, and K. R. Jennings. Employee participation: diverse forms and different outcomes. *Academy of Management Review* 13: 8–22, 1989.

Dachler, H. P. and B. Wilpert. Conceptual dimensions and boundaries of participation in organizations: a critical evaluation. *Administrative Science Quarterly* 23: 1–39, 1990.

Elias, J. L. and S. Merriam. *Philosophical Foundations of Adult Education.* Malabar, FL: Robert E. Krieger, 1980.

ILO/WHO Committee on Occupational Health. *Psychosocial Factors at Work.* Geneva: ILO, 1986.

King, P. The psychosocial work environment: implications for work-place safety and health. *Professional Safety,* March 1995.

Ladou, J. In *Occupational Safety and Health.* Chicago: National Safety Council, 1987.

Likert, R. *The Human Organization.* New York: McGraw-Hill, 1967.

Levi, L. *Stress in Industry.* Geneva: ILO, 1984.

Lischeron, J. Unpublished paper. Calgary, 1984.

Wallerstein, N. and M. Weinger. Health and safety education to worker empowerment. *American Journal of Industrial Medicine* 22: 619–635, 1992.

PART IV | *Decision to Err*

Chapter *13*

Logical Decision to Err

The decision to err, or to work unsafely, is often a very logical choice that is made either consciously or unconsciously by the worker. In a particular work situation it does not make sense to the worker to perform the job safely as instructed. This fact is not yet well accepted in the field of safety. It does, however, make a great deal of sense in the light of behavioral science theory and experimentation.

There are three aspects of decision to err: (1) due to the worker's psychological situation, he or she chooses the only logical behavior open to him or her—to work unsafely; (2) the worker unconsciously chooses unsafe procedures to satisfy some unknown need; and (3) the worker simply chooses the unsafe procedure because he or she does not think anything will happen to him or her.

The motivational field of the worker is depicted in Figure 11-1. In that model, performance is depicted as a function of two things, the worker's ability and his or her motivation. The model suggests that after performance, the worker's decision whether or not to perform again depends on the reward received for that performance (both extrinsic and intrinsic) and the worker's perception of that reward compared to what he or she thinks would be a fair reward for that performance (satisfaction). This process is subconscious, but it does occur. The worker's level of satisfaction is an important determinant of his or her future behavior.

The ability component of performance, according to the model in Figure 11-1, is a function of selection and training. The motivational component (far more important in determining behavior) is a function of a great many things, as indicated. In this chapter, we will concentrate on the peer group, personal factors, the boss's style, and the boss's measures and priorities. These are probably major determinants in worker decisions to work unsafely.

REINFORCEMENT AS A DETERMINANT

The model in Figure 13-1 proposes that future performance (behavior) is dependent on what happened as a result of past performance. This, of course, is standard learning theory from psychological texts, or, if you prefer, is the standard behavior modification theory proposed by Skinner. Reinforcement is accepted universally as a determinant of performance; it does determine behavior.

An employee has two choices: desired performance or undesired performance (safe or unsafe act). If the employee performs as desired and is rewarded in some way, he or she will be more likely to repeat that desired performance. If the employee does not perform as desired and, as a result, is penalized in some way, he or she will be less likely to repeat that undesired performance. Both situations tend to build safety behavior (desired performance). Actually, the first situation builds safe behavior, and the second tends to make unsafe behavior less likely; both outcomes are desirable.

If, however, the employee performs as desired and is penalized, he or she will be less likely to repeat that desired performance. If the employee does not perform as desired and is rewarded, he or she is likely to continue to repeat that undesired performance. These two situations lead to the building of error-prone performance. As indicated in Figure 13-1, favorable and unfavorable consequences (or reinforce-

Employee			
Desired performance		Undesired performance	
Favorable consequences	Unfavorable consequences	Favorable consequences	Unfavorable consequences
Money	Money	Money	Money
Worker-boss-relationship	Worker-boss relationship	Worker-boss relationship	Worker-boss relationship
Coworker relationships	Coworker relationships	Coworker relationships	Coworker relationships
Customer relationships	Customer relationships	Customer relationships	Customer relationships
Other	Other	Other	Other
←		Error prone	→

Figure 13-1. Reinforcement of desired or undesired performance. (Adapted from M. Moore, Behavioral change and behavioral technology, *ASTME Vectors,* Vol. 5, 1969.)

Logical Decision to Err

ments, or rewards) are money, good worker–boss relationships, good coworker relationships, good customer relationships, and so on.

One of the reasons why people commit unsafe acts (undesired performance) is that they have been rewarded in the past for doing just that, and/or they have been ignored or penalized in the past for working safely. If management has ignored (or penalized) workers for safe behavior, or rewarded (in the workers' eyes) unsafe behavior, management has made unsafe behavior a logical choice for workers.

Another model is shown in Figure 13-2. This individual-performance model suggests that external behavior is influenced by a number of cues coming from the boss and the peer group, and by rewards coming from the boss and

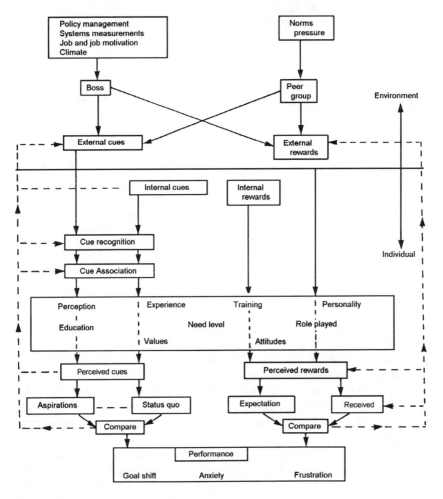

Figure 13-2. An individual-performance model.

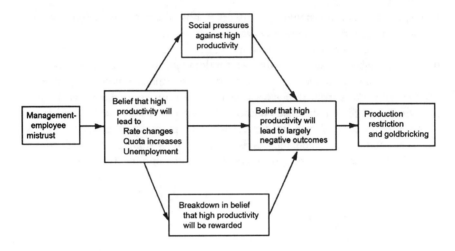

Figure 13-3. Causes of production restriction and goldbricking. (From A. L. Porter, E. Lawler, and J. Hackman, *Behavior in Organizations.* New York: McGraw-Hill, 1975. Copyright © 1975. Used with permission of McGraw-Hill Book Company.)

the peer group. These cues are interpreted internally and are combined with other internal cues, telling the worker how to act in the situation he or she is in. The external rewards are combined with internal rewards (how the worker feels about the performance). These cues and rewards are filtered through the worker's normal filtering process (a combination of his or her perception, education, experience, value system, need level, training, attitude, role, and personality). Filtered (perceived) cues lead the worker to a comparison of what he or she wants (aspirations) and what he or she has (status quo). Filtered (perceived) rewards lead the worker to a comparison of what he or she expected to get to what he or she got. These comparisons lead to either performance or lack of performance, which is accompanied by anxiety, frustration, or a change in goals.

A third model, shown in Figure 13-3, was developed to study goldbricking, but it seems relevant to safety also. Our inability to elicit desired performance (safe behavior) might well be a function of a belief that safe performance will lead to largely negative outcomes because of social pressure, the belief that high productivity is rewarded more than safety, and so on. All of this relates to management–employee mistrust.

In short, employee behavior is a function of what rewards and reinforcements have been received in the past for past behaviors. These rewards are both intrinsic and extrinsic, and the primary extrinsic rewards have been through interpersonal relationships with the boss or the peer group.

Logical Decision to Err 199

THE PEER GROUP

Each employee is not only an individual but also an integral part or member of a group. Each manager must manage his or her crew as individuals and also as a group. Every group has a distinct personality of its own, makes its own decisions, and sets its own work goals. These may be identical with management's goals, or they may be different. The group also sets its own safety standards and lives by these standards regardless of what standards management or OSHA may have set.

A group is a number of people who interact or communicate regularly and who see themselves as a unit distinct from other collections of people. Also, the members of a group are drawn, or perhaps even bound, to one another in one or another state of interdependence. In other words, something is at stake, and the group members share that something. This interdependence may have nothing to do with the task that the group performs, but does have to do with the group itself or with the relationships within it. Thus each member may depend on the others to satisfy the need for affection or affiliation or security. To be a group, the people in question must communicate or interact regularly, depend on one another, and think of themselves as a group.

The influence of the group on the individual group member is shown in Figures 13-4 and 13-5. The group influences the individual's personal characteristics

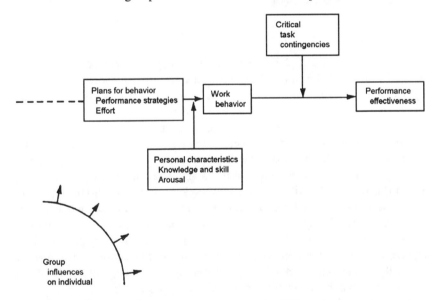

Figure 13-4. Major determinants of group-member behavior and performance effectiveness. (From J. Hackman, Group influences on individuals, in *Handbook of Industrial and Organizational Psychology,* edited by M. Dunnette. Chicago: Rand-McNally, 1976.)

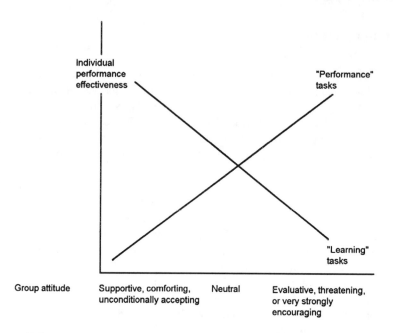

Figure 13-5. The individual's relationship with the group. (Adapted from J. Hackman. Group influences on individuals, in *Handbook of Industrial and Organizational Psychology,* edited by M. Dunnette. Chicago: Rand-McNally, 1976.)

(knowledge, arousal) and plans for behavior (strategics, effort). In short, group influence is a major determinant of an individual's work behavior.

A factor that often influences the safety behavior of a person is what sociologists call *group norms*. Group norms are really the informal laws that govern the way the people that belong to a group should and should not behave. Very often, when members of a group are asked what their norms are, they cannot identify them; yet unconsciously their behavior is influenced by them.

Group norms are the accepted attitudes about various things in the group situation. They include attitudes about how workers behave toward the boss, how they react to safety regulations, and how they react to production quotas. Group norms are the codification of members' attitudes about the company, manner of dress, and merit systems into recognized, accepted, and enforceable behavioral patterns. If a member of a group takes on a pattern of behavior or expresses an attitude that is in violation of that commonly accepted by the group, the group has ways of punishing the member to bring him or her back into line.

In an industrial organization, if the norms developed within the work group are favorable to safety, the group itself will encourage and even enforce safe practices much better than supervisors can (the wearing of hard hats by construction workers, for instance). Group norms often develop that are against safety rules; for example, a group of workers might have the attitude that safety is a waste of their time.

Often, management's first response to group disregard of safety rules is to pass a regulation that will force the individual to violate the norms of his or her peers and comply with the safety rule. If the group is a strong group (with a high degree of cohesiveness), the member will violate management's directive rather than run the risk of being excluded from the work group. Our objective should be to find some way to change the group norm and get this phenomenon working for safety, rather than against it.

Group pressures put more pressure on workers than do the standard procedures written in the safety manuals. Regarding safety, work-group pressures and group norms are perhaps the most important determinants of worker behavior. To reiterate: the group sets its own safety rules, and its members live by those rules, not management's.

The safety program, then, must not only speak to the individual; it must also attempt to understand the group's safety norms and to influence those group norms so that they are safety-oriented norms. The safety program must help to build strong work groups with goals that coincide with safety goals.

Group pressures are not exerted like management pressures; groups do not issue orders and provide training. Rather, the group member observes what is going on in the group and adopts its norms and behaviors because this makes the member feel more secure. Also group thoughts and actions may seem more "right" to the individual than his or her own thoughts and actions. Individuals do yield to group pressure regarding opinions, attitudes, and even self-evaluation.

Figure 13-5 shows that, as regards job performance (performance tasks), the more evaluative, threatening, or strongly encouraging the group is, the better the individual performs as the group wishes. Similar pressure from the group, however, makes learning more difficult.

Group Effectiveness

Figure 13-6 shows some of the factors that contribute to group effectiveness and strength. Our concern is with influencing the independent and intermediate variables in order to maximize the dependent variables. Most of the independent variables in the chart are controlled to some degree by a manager, who may be able to choose the membership of the group or to rearrange the communication network. The manager might also be able to control the size of the group, its physical setting, and the nature of its task. The strength of a group can be determined by observing some of its characteristics. In a strong group, the members voluntarily:

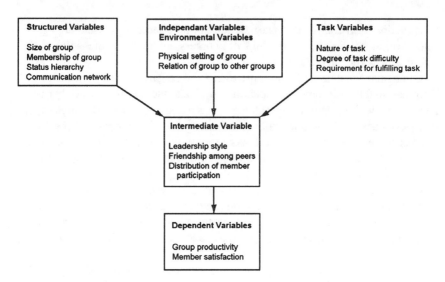

Figure 13-6. Factors that contribute to group effectiveness. (Adapted from D. Krech, S. Crutchfield, and E. Ballachey, *Individual in Society.* New York: McGraw-Hill, 1962.)

- Try to deserve praise from the rest of the group.
- Seek recognition from the group leaders (not management).
- Exert pressure on weak group members.
- Put special efforts into achieving the group goals.

The key to identifying strong groups seems to lie in the word *voluntarily.* In a strong group, the members seem to desire all four achievements. It is important to them as individuals to conform to the goals and mores of the group.

In a weak group, the members:

- Form cliques or subgroups.
- Exhibit little cooperation.
- Are unfriendly.
- Use no initiative.
- Avoid responsibility.
- Have no respect for company policies.

Some causes of weak-group cohesiveness are too much pressure from management, lack of support, and job inadequacy.

Each manager might do well to stand back and observe his or her group. If the manager observes any of the characteristics of a weak group, perhaps he or she needs to try to build a stronger group. To build a strong group with goals identical to management's, the manager must first strive to build a strong group. Four things are essential for a strong group:

1. Individual competence. Each member of the group must have the ability to pull his or her own weight.
2. Individual maturity. Each member must be mature. This means that the group dislikes the "prima donna" who can but who will not do his or her share. The group dislikes the "yes" person who exerts more effort pleasing the boss than the group. It also dislikes the "let George do it" type.
3. Individual strength. Each member must have not only the ability and the maturity but also the strength to do his or her job to earn group respect. This means that weaklings and loners are not tolerated.
4. Common objectives. These are essential to a strong group. Table 13-1 provides a procedure for rating group effectiveness.

Building Stronger Groups

In the literature, there is a classic demonstration of the social and psychological forces that can build group cohesiveness. A group that lacked cohesiveness was transformed into a highly cooperative team within a few days. This quick building of cohesiveness was not accomplished with requests for teamwork; no platitudes about cooperation, slogans, or posters picturing the one weak link in the chain of cooperation were used. The scientists merely used the natural forces that are inherent in groups.

The subjects of the experiment were 12-year-old boys at a secluded camp provided by the Yale University Department of Psychology. The boys did not realize that they were part of an experiment, and the man they thought was the camp caretaker was Muzafer Sherif, the chief psychologist. The following is a list of the procedures used and the conditions created that built a random collection of boys into two highly cohesive groups:

1. Physical proximity. The boys were strangers to one another at the start. After they had been in camp a few days, they were divided into two groups. Each group had separate living quarters. This is similar to most business situations, in which strangers are put together in a room separated from other work groups. Being thrown together physically gave the boys opportunities for interactions that would not otherwise have taken place, thus fulfilling the first criterion for building team spirit, at least on a "departmental" level.
2. Sharing of common goals. The two separate groups of boys proceeded to set goals for their respective groups. They decided on the decoration and the arrangement of their quarters, and chose activities that appealed to

Table 13-1. Rating group effectiveness

Goals		
Poor	1 2 3 4 5 6 7 8 9 10	Good
Confused; diverse; conflicting; indifferent; little interest		Clear to all; shared by all; all care about the goals; feel involved
Participation		
Poor	1 2 3 4 5 6 7 8 9 10	Good
Few dominate; some passive; some not listened to; several talk at once or interrupt		All get in; all are really listened to
Feelings		
Poor	1 2 3 4 5 6 7 8 9 10	Good
Unexpected; ignored or criticized		Freely expressed; emphatic responses
Diagnosis of Group Problems		
Poor	1 2 3 4 5 6 7 8 9 10	Good
Jump directly to remedial proposals; treat symptoms rather than basic causes		When problems arise the situation is carefully diagnosed before action is proposed; remedies attack basic causes
Leadership		
Poor	1 2 3 4 5 6 7 8 9 10	Good
Group needs for leadership not met; group depends too much on single person or on a few persons		As needs for leadership arise various members meet them; anyone feels free to volunteer as he or she sees a group need
Decisions		
Poor	1 2 3 4 5 6 7 8 9 10	Good
Needed decisions don't get made; decisions made by part of group; others uncommitted		Consensus sought and tested; deviates appreciated and used to improve decision; decisions when made are fully supported
Trust		
Poor	1 2 3 4 5 6 7 8 9 10	Good
Members distrust one another; are polite, careful, closed, guarded; they listen superficially but inwardly reject what others say; are afraid to criticize or to be criticized		Members trust one another; they reveal to group what they would be reluctant to expose to others; they respect and use the responses they get; they can freely express negative reactions without fearing reprisal
Creativity and Growth		
Poor	1 2 3 4 5 6 7 8 9 10	Good
Members and group in a rut; operate routinely; persons stereotyped and rigid in their roles; no progress		Group flexible; seeks new and better ways; individuals changing and growing; creative; individually supported

Source: Adapted from C. Handy, *Understanding Organizations.* Middlesex, Eng.: Penguin Books, 1976.

them. Each group worked toward these goals independently. Sharing decision making and then working together to reach shared goals are prime factors in cohesiveness building.

3. Setting up of an organization and acceptance of leadership. These boys had not worked together more than a few hours before they began to pool their efforts. They spontaneously organized duties within the groups. They noticed that some members were adept at special activities, so they

created positions in their organizational charts for these experts. They quickly divided the work and defined the responsibilities of different members. Each member soon understood what role he was expected to play in group life. These groups also quickly came to look to a few members to play "higher roles" in coordinating the others. Captains and lieutenants emerged, and group activities began to center around them. The groups set up their own social levels, or hierarchies of power. But their accepted leaders were from within the group, not from outside it. In a business setting, the organizational hierarchy already exists when a new member enters. The person the company designates as boss may not be the one the group would have designated. Thus organizations usually have appointed leaders while groups have their informal leaders.

4. Development of group symbols. The boys had scarcely agreed on their accepted leaders before the members were clamoring for symbols to identify themselves as distinct groups. They invented nicknames and some jargon for their activities. Industrial groups do this if they are cohesive, and the vocabulary of one department may sound like Greek to the department down the line. The boys also developed some secrets, as offices do through the grapevines, and as families do in family jokes.

The boys' groups bought caps and T-shirts, in the symbolic colors decided on. Adults seem to have much of this same "kid stuff" in them. Railroad workers favor a certain style of work clothing that is a trademark of their group. A house painter feels disloyal to his or her occupational group unless he or she works in painter's whites. The blue shirt and the white shirt are group symbols in the business hierarchy. The work clothes suit the roles that members are expected to play. When groups want such distinctive symbols, it is evidence of cohesiveness. Work uniforms not agreed on by the group do not necessarily build cohesiveness, however.

5. Competition with natural enemies. Each group of boys quickly looked upon the other group as a natural enemy. Groups tend to hold together more firmly when threatened by some enemy, or when some stress makes the members realize they are dependent on cooperation for security or perhaps for survival. Rivalry and stressful situations are not rare within a business. One department often looks on another as a natural enemy. One clique considers the other clique a rival, and each clique then holds together more strongly than before, and cooperates less and feuds more with the rival clique. If the appointed leader is dogmatic and self-centered, the workers may become more cohesive, but will cohere around the goal of frustrating the boss rather than cooperating with him or her.

In the case of the boys' groups, the experimenters exploited natural rivalry by egging the two groups into competitive contests. Little encouragement was needed; each group was itching to prove its superiority. To intensify this rivalry, the experimenters rigged some of the contests. This made the losers furious at their opponents. Each group held closer together than ever and engaged in open as well as secret warfare. There were pitched battles in which the boy who had previously been the cry-baby overnight became hero of his group and a despised villain to the other group. To protect life and limb, it became necessary for the experimenters to order the hostilities stopped. Merely giving the order and policing the groups was not even adequate at this point.

6. Friendship among group members. Many social psychologists think this item is the most important factor in building cohesiveness. A person has got to like the people to become a part of the group. There is, however, a reciprocal relationship necessary here. Cohesiveness seems to be built most easily when the people are mutually attracted to begin with. But as cohesiveness does develop, people who have not previously seemed attractive become so if they are in the group. The people in a group usually seem to other group members to be a little more capable and attractive than their counterparts in competing groups.

Research and experimentation suggest some ways to build stronger work groups. We can use groups to facilitate safety programs if we:

1. Place people in physical proximity. If the workplaces do not automatically allow this (as in a shipping department), we can bring people together periodically for five-minute safety talks or decision-making meetings.
2. Allow the group to set its own safety goals.
3. Allow the group to organize its safety procedures. Let it select its own "enforcers," inspectors, talk givers, committee representatives, departmental monthly safety directors, and so on.
4. Allow the group to develop its own symbols, safety nicknames, and so on.
5. Set up competitions with rival departments as regards safety records, inspection results, sampling results, and so on.
6. Allow departments or groups some latitude in selecting their own members. Or at least allow transfers if department members do not seem to get along well.

ATTITUDES

As reported by Brayfield and Crockett, the classic study relating attitudes to performance in an industrial setting was conducted in 1930 in Neenah, Wisconsin, in a mill operated by Kimberly-Clark Corporation. Between 200 and 300 young girls engaged in routine repetitive jobs at machines were administered questionnaires and interviews. The researchers found that there was no relationship between attitudes and performance. They did find that unfavorable attitudes correlated slightly with time lost because of sickness. Similar findings were obtained by Brayfield in 1944 among office workers, and in 1950 among plumbers' apprentices. Similar studies have been done with farmers, air force control-tower operators, and IBM operators. Correlations have been found between job satisfaction and productivity for life insurance salespersons, retail sales clerks, and hourly workers in the aircraft industry. Most of the studies of individual workers indicate no correlation between job satisfaction and job performance. Studies of groups of workers, however, do show evidence of a relationship between job satisfaction and performance. Group studies are performed by using an attitude survey and then correlating that to measures of the group's work effectiveness.

Attitudes have been studied as they relate to other indicators important to management, such as absenteeism and employment stability. There seems to be some relationship, but it is by no means clearly established.

Several studies have been done on the relation of attitudes to accident records (Figure 13-3 being one example). Stagner, Flebbe, and Wood used a group-analysis design to study the job satisfaction of railroad employees in ten divisional groups. The correlations obtained between job satisfaction and the accident-record were negative and small. A study by Yoder, Henemen, and Cheit found that employees in three firms with fewer-than-average accidents had a mean attitude score of 133, while employees in three firms with more-than-average accidents had a mean score of 143. Although the data were limited, the researchers concluded that there is a tendency for firms with higher accident rates to have more favorable attitude scores (143).

Attitude Development

Robert Mager's book, *Developing an Attitude Toward Learning,* gives us an insight into attitude development that can be directly applied to developing safety attitudes. He reported the findings of a study he made to determine students' attitudes toward different academic subjects and what formed the attitudes. According to Mager's study, a subject area tends to be favored because the person seems to do well at it, because the subject is associated with liked or admired friends, relatives, or instructors, and because the person is relatively comfortable when dealing with the subject. Conversely, a least-favored subject seems to become that because of a low aptitude for it, because it is associated with disliked individuals, and because the subject matter is associated with unpleasant conditions.

Thus, the main factors that help mold an attitude toward a subject are: the conditions that surround it, the consequences of coming into contact with it, and the way that others react toward it (modeling).

Conditions

When a person receives instruction in significant subject matter (safety, for instance), he or she should be in the presence of as many positive and as few negative conditions as possible. If a subject that initially has no special significance is presented to someone on several occasions while he or she is undergoing unpleasant experiences, that subject may become a signal to escape, to get away from the unpleasantness. On the other hand, if a person is introduced to a subject under pleasant conditions, that subject may become a signal to stick around because the person likes the association.

Mager illustrated this concept by describing the reaction of most people when a doctor moves a hypodermic needle toward them. They tend to back away or turn their heads to avoid seeing this signal of forthcoming pain. There is nothing bad about the sight of the hypodermic needle the first time we see one. But after experiencing pain in the presence of the needle, the sight of it becomes a signal of coming pain. It is as though the mere sight of the needle becomes a condition to be avoided.

Consequences

How a supervisor reacts to a worker's efforts to learn about and follow safety procedures is another important factor in determining the success or failure of a safety program. If you want to increase the probability that a response will be repeated, follow it immediately with a positive consequence.

If you want to reduce the probability that the behavior will occur again, follow it immediately with an unpleasant (aversive) consequence.

Aversives and Positives

What in a work situation is an aversive, and what is a positive? Although it is not always possible to know whether an event is positive or aversive for a given individual, some conditions and consequences are universal enough to provide us some direction.

First consider the aversives. Mager suggested that we define an aversive as any condition or consequence that causes a person to feel smaller or makes his or her work seem inconsequential. Here are some common aversives:

  Pain
  Fear and anxiety
  Frustration
  Humiliation

Embarrassment
Boredom
Physical discomfort

Positives create an atmosphere that is more pleasant for everyone—the supervisor and the workers. There is less tension and a greater willingness to consider new ideas. Here are some positive conditions or consequences:

Rewarding positively
Understanding where the worker is
Teaching only what the worker needs
Feedback
Involvement
Participation
Treating the worker as a person
Giving the worker responsibility

Modeling

Another way in which behavior is strongly influenced and attitudes are formed is through modeling (learning by imitation). The research on modeling tells us that if we want to teach workers behaviors, we must exhibit those behaviors ourselves. In other words, we must behave the way we want our employees to behave.

THE BOSS'S STYLE, MEASURES, AND PRIORITIES

Corporate climate and managerial style seem to be closely interdependent, and to a large degree climate is dictated by style. It seems to take all kinds of managers to keep a company going. Each kind of leader creates a human climate that has specific effects on his or her followers. Although the number of different methods used by leaders must be enormous, it is easy to classify most leaders according to one or another style of leadership. The style classifications most commonly used are autocratic and democratic. There are many other classifications, most of which could also be described as democratic and autocratic. Table 13-2 lists some of these styles.

Rensis Likert described the relative effectiveness of the two basic styles as follows:

1. Supervisors with the best records of performance focus their primary attention on the human aspects of their subordinates' problems and on

Table 13-2. Some autocratic and democratic managerial styles

Autocratic Style	Democratic Style
Authoritarian	Egalitarian
Dictatorial	Facilitative
Leader-centered	Group-centered
Production-centered	Worker-centered
Restrictive	Permissive
Job-centered	Employee-centered

endeavoring to build effective work groups with high performance goals. These are referred to as employee-centered leaders.

2. The performance goals of supervisors are also important in affecting productivity. If a high level of performance is to be achieved, it appears to be necessary for a supervisor to be employee-centered and at the same time to have high performance goals. However, there is a marked inverse relationship between the average amount of "unreasonable" pressure the workers in a dependent position feel and the productivity of the department. Feeling a high degree of unreasonable pressure is associated with low performance.

3. General rather than close supervision is more often associated with a high rather than a low level of productivity. This relationship was found in a study of clerical workers for supervisors. Similar results were found for nonsupervisory employees. Supervisors in charge of low-producing units tend to spend more time with their subordinates than do the high-producing supervisors, but the time is broken into many short periods in which they give specific instructions: "Do this, do that, do it this way," etc.

4. Genuine interest on the part of a superior in the success and well-being of his or her subordinates has a marked effect on their performance. For example, high-producing foremen tend either to ignore the mistakes their subordinates make, knowing that they have learned from the experience, or to use these situations as educational experiences by showing how to do the job correctly. The foremen of the low-producing sections, on the other hand, tend to be critical and punitive when their subordinates make mistakes.

While Likert's work is aimed primarily at relating managerial style to production and not safety, he seems to have a message for safety managers. On the surface, at least, it would seem that employee-centered supervisors might generate better safety records than job-centered supervisors.

WHAT TO DO TO GET RESULTS

It would seem, on the basis of the many studies discussed in this chapter, that there are some things we can do to reduce the number of worker decisions to err. We have to make it more logical for an employee to choose not to err. We can foster this choice by: (1) positively reinforcing safe performance, making it more likely that the worker will repeat that safe performance in the future; (2) building strong work groups that have group norms that include safety; (3) building positive safety attitudes by creating positive, not aversive, conditions and consequences for safe behavior; and (4) encouraging bosses to be employee-centered and to make safety a high priority.

REFERENCES

Air Force Personnel and Training Research Center. *Reducing Traffic Accidents by Use of Group Discussion-Decision.* Randolph AFB, TX, 1957.

Andrews, T. *Methods of Psychology.* New York: Wiley, 1948.

Brayfield, A. and W. Crockett. Employee attitudes and employee performance. In *Readings in Organizational Behavior and Human Performance,* edited by L. Cummings and W. Scott. Homewood, IL: Irwin-Dorsey, 1969.

Brody, L. *Accidents and Attitudes.* New York: New York University, 1959.

Dunbar, F. Susceptibility to accidents. *Medical Clinics of North America,* Vol. 28, 1944.

Fulton, W. Industrial medical potentials. *Industrial Medicine and Surgery,* Vol. 18, 1949.

Hackman, J. Group influences on individuals. In *Handbook of Industrial and Organizational Psychology,* edited by M. Dunnette. Chicago: Rand-McNally, 1976.

Handy, C. *Understanding Organizations.* Middlesex, Eng.: Penguin Books, 1976.

Jenkins, T. Personality factors related to accidents by production employees. Mimeographed. New York: New York University, 1956.

Krech, D., S. Crutchfield, and E. Ballachey. *Individual in Society.* New York: McGraw-Hill, 1962.

Lauer, A. Comparison of group paper and pencil tests with certain psychological tests for measuring driving aptitude of army personnel. *Journal of Applied Psychology,* Vol. 39, 1955.

Levinson, H. The illogical logic of accident prevention. *Menninger Quarterly,* March 1957.

Likert, R. *New Patterns of Management.* New York: McGraw-Hill, 1961.
Mager, R. *Developing an Attitude toward Learning.* Belmont, CA: Fearon, 1968.
Moore, M. Behavioral change and behavioral technology. *ASTME Vectors,* Vol. 5, 1969.
Newcomb, T. The prediction of interpersonal attraction. *American Psychologist,* 575–586, 1956.
Petersen, D. The safety profession: concepts and programs, *Professional Safety,* December 1984.
Porter, A. L., E. Lawler, and J. Hackman. *Behavior in Organizations.* New York: McGraw-Hill, 1975.
Rainey, R. An investigation of the role of psychological factors in motor vehicle accidents. Unpublished paper, 1958.
Selling, L. Psychiatric findings in the cases of 500 traffic offenders and accident-prone drivers. *American Journal of Psychiatry* 68–79, 1940.
Sherif, M. Experiments in group conflict. *Scientific American* 54–58, 1956.
Skinner, B. *Science and Human Behavior.* New York: Macmillan, 1963.
Smith, T. Rebutting behaviorism, in *Industrial Safety and Hygiene,* March 1995.
Stagner, R., D. Flebbe, and R. Wood. Working on the railroad: a study of job satisfaction. *Personnel Psychology* 5:293–306, 1952.
Stiles, G. Relationships of unmet emotional needs to accident repeating tendencies in children. Ph. D. dissertation, New York University, 1957.
Tillman, W. and G. Hobbs. The accident prone automobile driver: a study of psychiatric and social background. *American Journal of Psychiatry* 321–331, 1949.
Whitehorn, J. The human personality and the development of mature individuals. Unpublished paper, 1965.
Wisely, H. Personal characteristics of commercial bus drivers related to accident proneness. Ph.D. dissertation, Northwestern University, 1947.
Wolf, H. and R. Pearson. Happy workers mean fewer injuries. In *Safety and Health,* June 1992.
Yoder, D., H. Henemen, and E. Cheit. *Triple Audit of Industrial Relations.* Minneapolis: University of Minnesota Press, 1951.

Chapter *14*

Proneness

WHAT IT IS

The second type of decision to err is the unconscious decision: the worker decides unconsciously that it makes more sense to operate unsafely than safely. He or she needs to err for some (usually unidentified) reason. The combined effects of the worker's state, environment, and job situation may cause the worker subconsciously to try to find an accident to become involved in. We used to call this *accident proneness;* and although the term is still used in some situations, the concept has been somewhat redefined.

First, there is a difference between an *accident repeater* and an accident-prone person. An accident repeater is an individual who has more than one accident of the same type. An accident-prone person has significantly more accidents than others. Research indicates that there is no such thing as one type of accident-prone person. Rather, some of the behavior of each individual is safe and some of it unsafe, depending on many things, including the environmental hazards to which he or she is exposed. American safety experts now believe that accident proneness exists in some people for at least short periods of time, exists in others for relatively long periods of time, and can be predicted for either type of person if the situation is assessed properly at the right time. If an individual has one or more accidents, it does not mean he or she is accident-prone. Accident proneness refers to relatively constant characteristics that make the person more susceptible to accidents. There are such people, but their number is small, and their contribution to the total accident problem is slight. Almost all people have accidents. When a person has difficulty adjusting to the environment, he or she is referred to as temporarily accident-prone or accident-susceptible.

Studies have tended to deemphasize the concept of accident proneness as a major cause of accidents. Schulzinger's survey of 27,000 industrial and 8,000 nonindustrial accidents indicated that the accident repeater was involved in only 0.5% of them, whereas 75% involved the relatively infrequent experiences of a large number of people. Schulzinger came to these conclusions:

1. The tendency to have accidents is a phenomenon that passes with age, decreasing steadily after reaching a peak at the age of 21. The accident rate between the ages of 20 and 24, in both industrial and nonindustrial studies, is two and one-half times higher than between the ages of 40 and 44, four times higher than between the ages of 50 and 54, and nine times higher than between the ages of 60 and 65.
2. Most accidents involve young workers. Of the nonindustrial accidents studied, 70% involved workers under the age of 35, and nearly 50% involved workers under 24.
3. Men are significantly more likely to have accidents than women; the ratio of male to female accidents was 2:1 in the nonindustrial studies and apparently even higher in the industrial studies.
4. Most accidents are due to the relatively infrequent, solitary experiences of large numbers of individuals (86%). These figures were identical for the industrial and the nonindustrial studies and remained constant for nearly every year of a 20-year period.
5. Those who suffer injuries every year during a three-year period (3–5%) account for a relatively small percentage of all the accidents (0.5%).
6. Irresponsible and maladjusted individuals are significantly more likely to have accidents than are responsible and normally adjusted individuals.

Simple chance is, of course, also a factor in the unequal distribution of accidents. A "normal" distribution of accidents (all random happenings) would show that a few people have many more accidents than most people have.

Schulzinger's studies indicated that when the period of observation is sufficiently long, it can be seen that most accidents occur in individuals with a low degree of proneness, and that the relatively small percentage of the population that has a disproportionately high number of accidents is essentially a shifting group, with new persons constantly entering and leaving. His experience suggests that in the course of a life span almost any normal individual under emotional strain or conflict may become temporarily accident-prone and suffer a series of accidents in fairly rapid succession. Most people, however, find solutions to their problems, develop defenses against their emotional conflicts, and drop out of the highly accident-prone group after a few hours, days, weeks, or months. Some persons remain highly accident-prone throughout life, with or without years of freedom from accidents. These are the truly accident-prone individuals. As we have seen, however, they have only a small percentage of all accidents.

Most theorists today believe in this specific definition of accident proneness: that certain individuals are "looking" for an accident to become involved in. According to Schulzinger, these people constitute about 0.5% of the working population.

Throughout the years, many researchers have used statistical approaches to prove or disprove accident proneness. Paul Schugsta has described some of them:[1]

> Farmer and Chambers (1939), who coined the term, accident proneness, state: "Previous statistical investigations have shown that industrial workers exposed to equal risks were unequal in their liability to sustain accidents, and that this unequal liability was a relatively stable phenomenon, manifesting itself in different periods of exposure in different kinds of accidents. . . ."
>
> Not all the articles published affirm the theory of accident proneness. In a paper by Mintz and Blum (1949), they attacked the theory on the following basis:
>
> 1. Since accidents can be considered to be "rare" events, the proper statistical model to be used in interpreting the occurrence of accidents and whether these happen with greater than chance frequency in a given situation is the statistical model known as the Poisson distribution.
> 2. According to the Poisson distribution, 9% of a given population should have 39% of the accidents, and 39½% should have 100% of the accidents.
> 3. Hence, if accident proneness is to be considered a reasonable explanation for any given distribution of accidents, then the distribution should be more extreme than these, e.g., significantly fewer than 9% should have 39% of the accidents or 9% of the population should have significantly more than 39% of the accidents, etc.

The predominance of present-day literature on this point confirms that accident proneness does exist as a personal trait in some individuals. We have seen several theories which attempt to determine the cause of this accident proneness. These theories, as we have seen, stretch from difficulties in the youth of a person to his motor and perceptual speeds, to his adventurousness and extroversion, to the role of his ego.

The following study was made at a local pharmaceutical and biological drug company. The approximate number of employees at the plant site is 3,000.

First, two tables will be shown for reference. They cover information from 1970 and 1971. These were the only years for which records were available.

---

1. Reprinted with permission from P. Schugsta, The theory of accident proneness and the role of the Poisson distribution, *ASSE Journal,* November 1973.

The column marked "Number of Injuries" includes lost time and no lost time accidents (Tables 14-1 and 14-2).

The Poisson Distribution is a statistical model that applies to a situation where the probability of occurrence of an event is extremely small, while the opportunity of occurrence is extremely large. In the past, the Poisson Distribution model has been the standard model used for predicting accident occurrences per time interval. As can be seen from examining both tables, the Poisson Distribution is a poor model, in this situation, for predicting injury rate.

Because of this poor "fit" of the Poisson Model, this knocks down the main support of the Mintz and Blum (1949) anti-accident proneness theory mentioned earlier in this paper.

To reiterate: "Hence, if accident proneness is to be considered a reasonable explanation for any given distribution of accidents, then the distribution should be more extreme than these, e.g., significantly fewer than 9% should have 39% of the accidents or 9% of the population should have significantly more than 39% of the accidents, etc."

As can be seen from this study, for 1971: 1) 35.2% of the people have 100% of the injuries. 2) 8.36% of the people had 51.3% of the injuries. For 1970: 1) 52.9% of the people had 100% of the injuries. 2) 10.4% of the people had 43.9% of the injuries.

Table 14-1. Proneness and Poisson distribution

Number of Injuries	Actual Number of People Who Had This Many Injuries	Predicted by Poisson Distribution	Predicted by Negative Binomial Distribution	Negative Binomial Distribution Probability
0	1140	932	1147	.4710
1	783	395	689	.2830
2	258	430	336	.1381
3	133	138	151	.0622
4	53	33	65	.0268
5	29	6	27	.0113
6	17	1	11	.0047
7	13	0	5	.0019
8	5	0	2	.0008
9	2	0	1	.0003
10	2	0	0	.0001
11	0	0	0	0
12	0	0	0	0
13	0	0	0	0
14	1	0	0	0

Source: P. Schugsta, The theory of accident proneness and the role of the Poisson distribution, *ASSE Journal,* November 1973.
Note: Average plant population for 1970, 2436. Total injuries for 1970, 2368. Average injury index per person, .9614.

Table 14-2. Proneness and Poisson distribution

Number of Injuries	Actual Number of People Who Had This Many Injuries	Predicted by Poisson Distribution	Predicted by Negative Binomial Distribution	Negative Binomial Distribution Probability
0	2074	1565	2074	.6471
1	607	1122	588	.1834
2	256	402	261	.0814
3	124	96	130	.0405
4	56	17	68	.0212
5	34	2	37	.0115
6	16	0	20	.0062
7	21	0	11	.0034
8	3	0	6	.0018
9	5	0	4	.0012
10	3	0	2	.0006
11	3	0	1	.0003
12	2	0	1	.0003
13	0	0	0	0
14	1	0	0	0

Source: P. Schugsta, The theory of accident proneness and the role of the Poisson distribution, *ASSE Journal,* November 1973.
Note: Average plant population for 1971, 3205. Total injuries for 1971, 2298. Average injury index per person, .717.

Even by separating out the three departments with the greatest number of accidents and a higher accident index per person, the Poisson Distribution Model still did not apply. Please note the following chart (Table 14-3).

The model that does fit the actual accident experience is the Negative Binomial Distribution Model. This does not mean that the Negative Binomial Model should replace the Poisson Model in all cases. Only that the Negative Binomial Model fits this actual case better than any other statistical model.

A possible explanation for the success or failure of the Poisson Model in a particular case may be the amount of accident proneness found within that particular group. Perhaps the Poisson Model is effective for the entire world population, in predicting the likelihood of an accident; but in a certain group, because of varying amounts of accident-prone people, the Poisson Model is not effective.

All of the above relates to the traditional concept of "proneness"—persons more susceptible to accidents because "that's the way they are." Perhaps, however, "proneness" has to do with human traits (see Chapter 9).

Table 14-3. Proneness and Poisson distribution

Number of Injuries	Actual Number of People Who Had This Many Injuries	Predicted by Poisson Distribution	Poisson Distribution Probability
0	517	459	.29019
1	621	567	.35906
2	212	351	.22214
3	119	145	.09162
4	48	45	.02834
5	26	11	.00701
6	17	2	.00144
7	11	.4	.00025
8	4	.05	.00003
9	2	0	0
10	2	0	0
11	0	0	0
12	0	0	0
13	0	0	0
14	1	0	0

Source: P. Schugsta, The theory of accident proneness and the Poisson distribution, *ASSE Journal,* November 1973.
Note: Average number of people, 1580. Total number of injuries, 1955. Average injury index per person, 1.237.

WHAT TO DO ABOUT IT

Dr. Leon Brody suggests that the following affect accident proneness:[2]

1. The psychology of safe behavior is no more and no less than the psychology of human behavior in general. Thus, people do not change character when they get behind the wheel of a car. Rather their flaws and quirks become more manifest or significant when reflected in the behavior of a ton or two of steel moving in an irregular pattern at X miles per hour.
2. Driving a motor vehicle, like other activities in our society, is essentially a social undertaking where co-operative behavior and a sense of social responsibility are predominantly needed. Attitudes such as these cannot be considered apart from personality patterns—and the latter are likely to be rooted in the earlier years of life.
3. A person has to have some measure of satisfaction in his need for affection, his need for personal security, and his need for personal significance in order to develop and carry on in a way that will enable him to enact social roles involving effective participation with others, as in driving.

---

2. Reprinted with permission from L. Brody, *Accidents and Attitudes,* New York, New York University, 1959.

4. Needless to say, then, the problem of effecting safe behavior is no exclusive concern of driver educators and safety educators—unless we consider every educator a safety educator. And probably we should. No single course of study will do the job. Nor will any single methodological formula. Foundations must be laid in the earliest grades . . . and in the home.
5. It is through group dynamics, that is, the interpersonal or social influences in such processes as group discussion, group discussion-decision, and role playing, that we have techniques equal to the difficult task of improving attitudes. But expert programming is essential.
6. Intelligent discipline also has an important part to play in promoting safe behavior through modification of attitudes, particularly when other methods apparently are inadequate in themselves or impractical. Student safety courts and state point systems can serve this purpose well.
7. As a matter of fact, educational programming should include a properly balanced variety of approaches, rather than rely on just one or two techniques. And given desirable attitudes through such education, the average individual may be expected to use his school-acquired knowledges and skills not only for his own safety but also for the protection of others. This, of course, is the ultimate objective of all safety programs.
8. Finally this philosophical note: no attitude toward safety is sound if it fosters an unhealthy fear of an activity—or if it means a lessening of the intelligent or calculated types of risk taking that have in the past produced, and will continue to produce, social programs and welfare.

Accident proneness has been studied and discussed for many years, with much conflict and confusion. It does seem to be a part of the accident-causation picture, although a relatively few people might be considered truly accident-prone. These few are accident-prone probably because of some life situation or emotional adjustment difficulty. These people are difficult to identify and perhaps more difficult to deal with, but they are a part of our accident problem. Whether or not we will ever have solid, valid devices to identify accident-prone workers is questionable. Whether identifying and dealing with them will be cost-effective is also questionable. We will, however, pay for their accidents on the job or on the highway, for accident-prone individuals constitute a small percentage of the work force and a larger percentage of our accident problem than their percentage would indicate.

REFERENCES

Arbous, A. The psychology of repeated accidents in industry. *American Journal of Psychiatrists* 98:99–101, 1971.

Brody, L. *Accidents and Attitudes*. New York: New York University, 1959.

Brody, L. Personal characteristics of chronic violators and accident repeaters. *Bulletin No. 152,* National Academy of Sciences. New York: National Research Council, 1957.

Buros, O. *The Sixth Mental Measurements Yearbook.* Highland Park, NJ: Gryphon Press, 1965.

Eno Foundation for Highway Traffic Control. Personal characteristics of traffic-accident repeaters. Saugatuck, CT: The Foundation, 1948.

Farmer, E. and E. Chambers. A psychological study of individual differences in accident rates. *Industrial Health Research Board Report No. 38.* London: Industrial Health Research Board, 1926.

Farmer, E. and E. Chambers. Tests for accident proneness. *Industrial Health Research Board Report No. 44.* London: Industrial Health Research Board, 1929.

Froghatt, P. and J. Smiley. The concept of accident proneness: a review. *British Journal of Industrial Medicine* 12:1–12, 1964.

Geller, E. S. The psychology of occupational safety, in *Industrial Safety and Hygiene News,* January 1992.

Graham, S. Are your employees dead tired?, in *Safety and Health,* March 1995.

Greenwood, M. and H. Woods. The incidence of industrial accidents. *Industrial Health Research Board Report No. 4.* London: Industrial Health Research Board, 1919.

Greenwood, M. and G. Yole. An inquiry into the nature of frequency distribution representative of multiple happenings with particular reference to the occurrence of multiple attack of disease or repeated accidents. *Journal of Royal Statistics* 83:255–279, 1920.

Grunert, K. Accident germ—accident disposition—accident. *Zeitschrift fur Experimentelle und Angewandte Psychologie* (Gottingen) 8(4):519–529, 1961.

Haight, F. Accident proneness: the history of an idea. Reprint No. 240. Berkeley: University of California Institute of Transportation and Engineering, 1964.

Heath, E. Test and measurements as applied to accident prevention situations. *ASSE Journal* 8:7–14, 1963.

Jenkins, T. The accident prone personality. *Personnel* 33:29–32, 1956.

Jenkins, T. Identifying the accident prone employee. *Personnel* 38:56–62, 1961.

Johnson, H. The detection and treatment of accident prone drivers. *Psychological Bulletin* 43:489–532, 1946.

Keehn, J. Accident tendency, avoidance learning and perceptual defense. *Australian Journal of Psychology* 13(2):157–169, 1961.

Kerr, W. Complementary theories of safety psychology. *Journal of Social Psychology* 454:3–9, 1957.

Kunce, J. and B. Worley. Interest patterns, accidents and disability. *Journal of Clinical Psychology* 22(1):105–107, 1966.

Marbe, K. Practical applications of statistics of repeated events, particularly to industrial accidents. *Journal of the Royal Statistical Society* 387–547, 1927.

Mintz, A. and M. Blum. A re-examination of the accident proneness concept. *Journal of Applied Psychology* 33:195–211, 1949.

Newbold, E. A contribution to the study of the human factor in the causation of accidents. *Industrial Health Research Board Report No. 34.* London: Industrial Health Research Board, 1926.

O'Leary, P. An assessment of the effectiveness of the Mann attitude. Ph.D. thesis, Michigan State University, 1971.

Petersen, D. The safety profession: concepts and programs, in *Professional Safety,* December 1992.

Robertson, Scott. Increased production, increased fatalities, in *New Steel,* October 1994.

Schugsta, P. The theory of accident proneness and the role of the Poisson distribution. *ASSE Journal,* November 1973.

Schulzinger, M. *Accident Syndrome.* Springfield, IL: C. C. Thomas, 1956.

Shaw, L. The practical use of projective personality tests as accident predictors. *Traffic Safety Research Review* 9:34, 1965.

Slocombe, C. and W. Bingham. Men who have accidents: individual differences among motormen and bus operators. *Personnel Journal* 6:251–257, 1927.

Thorndike, R. The human factor in accidents with special reference to aircraft accidents. USAF Report No. 1. Washington, DC: U.S. Government Printing Office, 1951.

Tillman, W. and G. Hobbs. The accident prone automobile driver: a study of psychiatric and social background. *American Journal of Psychiatry* 106:321–331, 1949.

Vilardo, F. *Historical Development of the Concept of Accident Proneness.* Chicago: National Safety Council, 1967. Mimeographed.

Chapter *15*

Perception of Risk

The third reason behind the worker's decision to err is that he or she may simply perceive that the probability of getting caught at it is low—the old feeling that "it can't happen to me." There are really two parts to this belief in low probability: (1) "it simply cannot or will not happen to me," and (2) "even if it should happen it simply could not be serious enough to be worth worrying about."

This belief, so prevalent in the American work force, is among our biggest problems. There is no safety engineer or manager anywhere who has not had to deal with the belief. It is present in all people and is universally recognized as one of our biggest headaches. Safety professionals have been trying for years to overcome this simple perception of the American worker.

Traditionally, American males have been brought up to believe that safety and manhood are mutually exclusive. Chaytor Mason of the University of Southern California describes the effect of this belief as follows:[1]

> Today, I would like to talk about one of man's behavior patterns—commonly accepted through his history—accepted without question by millions of men—even today—the concept that manhood and safety are not compatible.
>
> *Case Study*
>
> A group of twelve to thirteen-year-old boys are [sic] standing at the base of a high voltage electric power line. They are arguing loudly. They are daring

---

1. Reprinted by permission from C. Mason, Manhood versus safety, in *Directions in Safety*, edited by D. Weaver & T. Ferry, Springfield, IL, Thomas, 1975.

223

each other to climb to the top (secretly each one is afraid). Finally one twelve-year-old breaks from the group and climbs quickly to the top and waves triumphantly to the group below. As he waves his arm touches the 55,000 volt line. There is a flash. He shudders—and falls—he plunges to the ground dead!

Case Study

A group of teenagers are [*sic*] racing down a Swiss mountainside on skis. One eighteen-year-old youth is falling behind. He takes a short cut that no one else would dare. It is a well known danger area—rocky—very steep—he must get ahead of the other boys. He misjudges a turn and crashes into a jagged boulder at forty miles per hour. He dies!

Case Study

The air is still and calm in this small coastal town. Fog has shrouded the airfield all night. The first flight of the day, Flight 54, a Convair Metropolitan, lets down into the fog—once—twice—and on the third pass finds the runway. Each approach was made below company and published minimums. But a successful landing was made.

Five minutes later, another Convair, Flight 53, drones over the field. Radio contact is made with the Passenger Agent (who also serves as weather observer). Flight 53 is advised that weather is well below minimum but that Flight 54 is on the ground preparing to leave.

Flight 53 overhead makes a VOR approach and pulls up after going 100 feet below the 390 foot minimum. On the second approach 200 feet below minimum ground witnesses can see the vague outline of the airplane as it pulls up again.

On the third approach, there is a sudden orange glow in the fog to the east as Flight 53 explodes through the roofs of three beach cottages at an altitude of 30 feet. Forty-one people die.

What is the common problem in all of these accidents? The problem was SAFETY VERSUS MANHOOD.

Many of us act as though looking after ourselves is a sign of cowardice.

The twelve-year-old could not say "It looks dangerous as hell to me—I am not going to climb up there." SAFETY VERSUS MANHOOD.

The eighteen-year-old skier had to prove his skill and daring at any cost. He had to take a chance. SAFETY VERSUS MANHOOD.

The pilot of Flight 54, who landed safely, knew that the pilot of Flight 53 would continue to make lower approaches until he got in—so he took three chances and made it. SAFETY VERSUS MANHOOD.

The copilot on Flight 53 died without a word. He could not express his fear to his Captain as they went below minimums. He would have seemed less of a man. SAFETY VERSUS MANHOOD.

The problem of Safety Versus Manhood can thus, occur at any time—at any age—in any job—in any activity. (Mostly this problem occurs in the teens and early adulthood but it can also have a resurgence in the "Fatal 40's.")

SAFETY VERSUS MANHOOD is one of the reasons why younger people have so many accidents in cars, in boats, in air—the standouts. Instead we read about how the people from the extreme ends of society handle their lives. (If they tell us the whole truth.) Of course, it is just possible that they sometimes varnished their lives a bit with some platitudes and some socially acceptable information. We read what they think they should say, not what they really did think and say.

Let's take this refusal to wear safety equipment as an example. Let's get a little historical background. Men's games have often been a picture of men's needs just as children's games are a picture of [a] child's needs.

Baseball was born in the United States in Cooperstown, New York, in 1847. It was a barehand game played with a wooden bat and the same rock hard ball that we know today. It was a game in which there were many injuries. The broken hand was the sign of the baseball player. The broken hand and the broken fingers were testimony that here was a baseball player. There was no protection for the hands, or any other part of the body.

At long last in the year 1875, Charlie Waite of the New York Nine, having suffered his third broken hand in a recent game (Blood priority; it takes a severe accident to cause awareness of danger)—came on the field wearing a thin leather protective glove. The audience was astonished and outraged. Pillows flew. (There were no pop bottles in those days.) "Get that 'sister' off the field." "If you are afraid of getting hurt get out!" "Don't muff it, catch the ball." And Charlie, a great first baseman, left the field never to return to baseball. Five years passed before another man attempted to wear a glove onto a playing field.

In 1880, another player, a budding lawyer, endured the ridicule of the crowd and stayed, wearing a glove, and slowly the baseball glove was accepted as part of the equipment for the first baseman. But only the first baseman.

In the year 1885, a well-known catcher named Charlie Bennett, who had many a floating rib sunk by a fast pitch from the pitcher, walked onto the playing field looking a bit more fat than usual. Charlie was wise to the ways of the crowd. Under his coat, he had hidden a thin quilted chest protector—adopted from the fencing sport. By the third inning two of the batters had gotten the idea. As one stepped up to the plate, the other tore open Charlie's coat and he was exposed in his shame. There for all the audience to see was the chest protector. He too was laughed off the field.

It was as late as 1883 that Arthur Irwin as a result of a broken hand popularized the glove for the infielders. The pitchers steadfastly refused to have any truck with such cowardly safety appliances until the year 1900. Even then, it was a sign of weakness to wear one. There were many disfigured catchers until the year 1893 when Nig Cuppy finally devised one out of a fencing mask. And you may recall the number of fractured skulls there have been from bean balls until the hard hat was recently developed for batters. Of course, it was developed first for the children's Little League—they were not supposed to be men. They could dare to protect themselves.

Since we are talking about protective equipment in games, let us examine another game. This one is older than baseball by quite a bit—and infinitely more dangerous. Even so, there are still problems in wearing safety equipment. It is the old and honored game we call *War!*

In 1914, there was not the vaguest hint of protective head gear. The Tommies left their pubs and marched to the field of battle wearing their green uniform caps. The Poilus kissed their mademoiselles good-bye at the sidewalk cafes and marched into battle wearing their bright red Kepis. The Fritz's dropped their steins and hastened to the playing field wearing their spiked leather decorative helmets. And each of them was an excellent target for any sniper, or piece of shrapnel coming his way.

It was not until 1915 that the steel helmet was developed. When it was, it was laughingly dubbed the "Tin Hat" and refused by many. Frankly, it was not the height of fashion to wear such a piece of head gear. In fact, one British General was so moved by the cowardice of the new breed of troops that he issued an order to his Division: "Any soldier in my command found wearing a "Tin Hat" will be court-martialed for cowardice in the face of the enemy." His order stuck and his division wore cloth hats for the rest of the war.

By the time the United States joined the big game in 1917, steel helmets were mostly worn, the style was established beyond question (although not without an occasional argument) and the American troops returned to Layfayette-land dressed to kill in "Tin Hats."

But there were exceptions. A senior officer, famous even then, refused to wear a steel helmet. In 1917, Colonel Douglas A. MacArthur stated to his aide, "I will not appear as a coward to my men by wearing a steel helmet." He hardly ever did. There was one exception though. In that same year, when he was to receive his dozenth Silver Star from General Black Jack Pershing, Pershing insisted that he wear a steel helmet, not because of the danger of the event, but only because all of the other Colonels were wearing them. That was one of the few times that MacArthur had ever been seen in a helmet. But how about Black Jack himself? There is no record that he *ever* wore one.

These examples point out two deadly beliefs prevalent among workers as well as sports heroes and national heroes: (1) that manhood and safety are natural enemies, and (2) that the incident simply will not happen to the person involved. These two beliefs are firmly ingrained, and perception is altered as a result of them.

Common sense in former times dictated that soldiers go into combat without hard hats and that baseball players play without gloves. People's perceptions at the time were that these behaviors were correct.

COMMON SENSE AND PERCEPTION

Communication theorists seem to be providing the best insights into the areas of common sense and perception, as evidenced by the words of Hans Toch and Malcolm MacLean:[2]

> Of most interest is common sense: Perception viewed through the eyes of common sense is clearly a passive affair. The eye is the equivalent of a motion picture camera, and hearing functions in the fashion of a tape recorder. The chemical senses act in the manner of variegated litmus paper; the mechanical senses register physical weights and measures. In other words, perception unassumingly transcribes on the slate of our awareness whatever the world presents to us. It dispassionately and uncritically records the gamut of bewildering impressions which reach us—mostly from without, but sometimes from within. This information, having been duly recorded, is then sorted, edited, and evaluated subsequently and—very importantly—elsewhere.
>
> In due fairness, one must add that common sense, when passed, may admit that there is probably more to the story. The senses, for example, don't appear to receive impressions at random: the eyes must be directed at some portion of the world, and the glass of wine must be sipped before anything of consequence is perceived in either case. Moreover, there is obviously some measure of control over the quality of the product: the languid gaze, the shameless stare, and the vacant look don't transmit comparable data. Sophisticated common sense also discovers that there is some question as to whether we always perceive equally well. Assuming, for example, that the cochlea responds with the same precision when a person sits in a concert hall or in his living room immersed in his newspaper, everyone knows that auditory awareness clearly differs in these situations.
>
> These and other observations of perception in action may suggest to common sense that the process is not altogether passive nor invariant. Perception seems to provide, within limits, the type of information the perceiver

---

2. Reprinted by permission from H. Toch & M. MacLean, Perception and communication, *Audio-Visual Communication Review* 10:55–77, 1967.

needs. Perception, in other words, is invoked, suppressed, and modified in the context of what the rest of the person is about. In order to be instrumental in this fashion, perception must be flexible and active. The vocabulary is full of words which imply recognition of this truism. The eye, for example, does not merely mirror or transmit; it scans, peeks, watches, stares, scrutinizes, and inspects. Such terms reflect a recognition of directionality, selection, or variability in perception.

Thus perception itself is not what is actually there, but rather what the person thinks is there. Everything is filtered through personal attitudes, values, experience, education, and so on. Similarly, what is common sense to one is not to another. Our safety rules are common sense to us, for we made them up. Our safety rules are often perceived as ludicrous to the worker, for they are not the kind of information that he or she needs. They may even be in opposition to the worker's group norms and thus in opposition to what he or she can accept. As Chaytor Mason indicated, people often go to great lengths to teach young men to be manly, to ignore danger, and then cannot understand why they ignore safety rules that are obviously good for them.

If we are going to change workers so that they better appreciate the fact that "it can happen to them," we have to do more than teach them, preach at them, and show them how incorrect their thinking is. These approaches, so traditional in safety programs, simply do not work. If we perceive safety differently from workers (and we usually do), we will not change their perceptions by teaching, preaching, and showing them the error of their ways.

The existence of perceptual differences is well documented by many. Toch and MacLean wrote:[3]

> The more complex a perceptual situation becomes, the greater the tendency for variations in perception to occur. Whereas a chair, for instance, provides a minimum of opportunity for differences in perception—at least, for members of our Western culture—any standard social situation constitutes a veritable perceptual cafeteria. This is the case not only because complexity multiplies the opportunity for the perceiver to assign meanings—for instance, one can choose to attend to one of many aspects of a complex situation in preference to others—but also because complexity usually evokes a wide gamut of personal experiences and needs which enter into the assignment of meaning.
>
> Hastorf and Cantril illustrate this process in their study of the football game between Dartmouth and Princeton which took place on November 23, 1951. The events which occurred in this game are conservatively catalogued as follows:

---

3. Ibid.

A few minutes after the opening kick-off, it became apparent that the game was going to be a rough one. The referees were kept busy blowing their whistles and penalizing both sides. In the second quarter, Princeton's star left the game with a broken leg. Tempers flared both during and after the game. The official statistics of the game, which Princeton won, showed that Dartmouth was penalized 70 years, Princeton 25, not counting more than a few plays in which both sides were penalized.

The sequel of these events was a prolonged and intense exchange of recriminations between players, students, coaches, administrative officials, student publications, alumni and partisans of the two universities, each of whom claimed to have sustained the brunt of the injuries.

Hastorf and Cantril submitted a questionnaire concerning the game to both Princeton and Dartmouth students and alumni, the results of which confirmed the divergent position of the two sides relating to the game. A film of the game also was shown to some 100 students; it yielded widely discrepant reports of the number of infractions committed by each side and the seriousness of these infractions. The Princeton students, for instance, "saw" the Dartmouth team make more than twice the number of infractions "seen" by Dartmouth students in watching the same film. They also "saw" two "flagrant" to each "mild" infraction for the Dartmouth team, and one "flagrant" to three "mild" offenses for their own team, a ratio considerably dissimilar to that of ratings by Dartmouth students. Hastorf and Cantril concluded:

> ... the "same" sensory impingements emanating from the football field, transmitted through the visual mechanism to the brain, obviously gave rise to different experiences in different people. The significances assumed by different happenings for different people depend in large part on the purposes people bring to the occasion and the assumptions they have of the purpose and probable behavior of other people involved.
>
> It is inaccurate and misleading to say that different people have different "attitudes" concerning the same "thing." For the "thing" simply is not the same for different people whether the "thing" is a football game, a presidential candidate, Communism, or spinach. We do not simply "react to" a happening or to some impingement from the environment in a determined way (except in behavior that has become reflexive or habitual). We behave according to what we bring to the occasion, and what each of us brings to the occasion is more or less unique.

The question is, of course, how do the workers see this "thing" called "safety"? We must understand their perception of safety before we ever can convince them that the probability of incident is higher than they think.

Perhaps the starting point in seeing it their way is to understand the motivational field that they live in at work. In Chapter 11, a model was presented depicting this motivational field. These motivators are the major influences coloring their perception; the primary "pulls" on them that develop their attitudes and determine their behavior.

ATTITUDES

Daniel Katz provides this definition of attitudes:[4]

> Attitude is the predisposition of the individual to evaluate some symbol or object or aspect of his world in a favorable or unfavorable manner. Opinion is the verbal expression of an attitude, but attitudes can also be expressed in nonverbal behavior. Attitudes include both the affective, or feeling core of liking or disliking, and the cognitive, or belief, elements which describe the object of the attitude, its characteristics, and its relations to other objects. All attitudes thus include beliefs, but not all beliefs are attitudes. When specific attitudes are organized into a hierarchical structure, they comprise value systems.
>
> The major functions which attitudes perform for the personality can be grouped according to their motivational basis as follows:
>
> 1. *The Adjustment Function.* Essentially this function is a recognition of the fact that people strive to maximize the rewards in their external environment and to minimize the penalties. The child develops favorable attitudes toward the objects in his world which are associated with the satisfactions of his needs and unfavorable attitudes toward objects which thwart him or punish him.
> 2. *The Ego-Defensive Function.* People not only seek to make the most of their external world and what it offers, but they also expend a great deal of their energy on living with themselves. The mechanisms by which the individual protects his ego from his own unacceptable impulses and from the knowledge of threatening forces from without, and the methods by

---

4. Reprinted with permission from D. Katz, The functional approach to the study of attitudes, *Public Opinion Quarterly* 24:163–204, 1960.

which he reduces his anxieties created by such problems, are known as mechanisms of ego defense.
3. *The Value-Expressive Function.* While many attitudes have the function of preventing the individual from revealing to himself and others his true nature, other attitudes have the function of giving positive expression to his central values and to the type of person he conceives himself to be. A man may consider himself to be an enlightened conservative or an internationalist or a liberal, and will hold attitudes which are the appropriate indication of his central values.
4. *The Knowledge Function.* Individuals not only acquire beliefs in the interest of satisfying various specific needs, they also seek knowledge to give meaning to what would otherwise be an unorganized chaotic universe. People need standards or frames of reference for understanding their world, and attitudes help to supply such standards.

Katz then provides us with some insight on attitude development (arousal) and attitude change:[5]

> The most general statement that can be made concerning attitude arousal is that it is dependent upon the excitation of some need in the individual, or some relevant cue in the environment. When a man grows hungry, he talks of food. Even when not hungry he may express favorable attitudes toward a preferred food if an external stimulus cues him. The ego-defensive person who hates foreigners will express such attitudes under conditions of increased anxiety or threat or when a foreigner is perceived to be getting out of place.
>
> The most general statement that can be made about the conditions conducive to attitude change is that the expression of the old attitude or its anticipated expression no longer gives satisfaction to its related need state. In other words, it no longer serves its function and the individual feels blocked or frustrated. Modifying an old attitude or replacing it with a new one is a process of learning, and learning always starts with a problem, or being thwarted in coping with a situation. Being blocked is a necessary, but not a sufficient condition for attitude change.
>
> Two conditions are relevant in changing value-expressive attitudes:
> 1. Some degree of dissatisfaction with one's self-concept or its associated values is the opening wedge for fundamental change. The complacent person, smugly satisfied with all aspects of himself, is immune to attempts to change his values. Dissatisfaction with the self can result from failures

---

5. Ibid.

or from the inadequacy of one's values in preserving a favorable image of oneself in a changing world.

To convert American prisoners of war, the Communists made a careful study of the vulnerability of their victims. They found additional weaknesses through a system of informers and created new insecurities by giving the men no social support for their old values. They manipulated group influences to support Communist values and exploited their ability to control behavior and all punishments and rewards in the situation. The direction of all their efforts, however, was to undermine old values and to supply new ones. The degree of their success has probably been exaggerated in the public prints, but from their point of view they did achieve some genuine gains. One estimate is that some 15 per cent of the returning prisoners of war were active collaborators, another 5 per cent resisters, and some 80 per cent "neutrals." Segal, in a study of a sample of 479 of these men, found that 12 per cent had to some degree accepted Communist ideology.

2. Dissatisfaction with old attitudes as inappropriate to one's values can also lead to change. In fact, people are much less likely to find their values uncongenial than they are to find some of their attitudes inappropriate to their values. The discomfort with one's old attitudes may stem from new experiences or from the suggestions of other people.

Managing Risk

Dr. Vernon Grose has been one of our leading experts on risk. His book on *Managing Risk* provides some good insights on this category of the causes of human error (risk taking):[6]

> Risk has been a component of life since the dawn of history. Moses wrote nearly 3,500 years ago: "Every new house must have a guard rail around the edge of the flat rooftop to prevent anyone from falling off and bringing guilt to both the house and its owner." Yet the good old days are filled with heroic epics that were fraught with risk: Columbus's first journey, Lincoln's decision to consider the United States indivisible, the settling of the western United States, the building of the Panama Canal, Lindbergh's New York-to-Paris flight, and even the relatively recent walks on the moon. All of these required decisions to take a chance. Isn't the statement, "worth the risk," a

---

6. Reprinted with permission from V. Grose, *Managing Risk,* Englewood Cliffs, NJ, Prentice-Hall, 1987.

Perception of Risk

tacit admission that opportunity was weighed against risk—that the risk was calculated?

Fundamental to trading among risks is the idea that risk is somehow tied to opportunity. Historically, this has been true. Figure 15-1 is a pictorial attempt to show this interrelationship. Note that two extreme tracks are shown: a *Prudent Path* and a *Gambling Grade*. Prudence is marked by a balance between risk and opportunity, covering a spectrum between security (where operational continuity is a high goal) and adventure (where growth potential is sought).

In contrast, gambling . . . exhibits a mismatch between risk and opportunity. At one extreme, opportunity far exceeds risk and typifies those rare instances—such as have occurred recently in computer technology—when explosive success results from an unbelievably low risk endeavor. The other end of the Gambling Grade is the inverse situation where risk totally outweighs opportunity, and only fools rush in there.

That zone where the two lines or courses of action intersect is classified as "ho-hum" because it suggests that the risk taker has failed to set any target or objective with respect to opportunity or risk.

To illustrate this link between risk and opportunity, recall the explosion of the space shuttle *Challenger*. Prior to that accident, NASA promoted oppor-

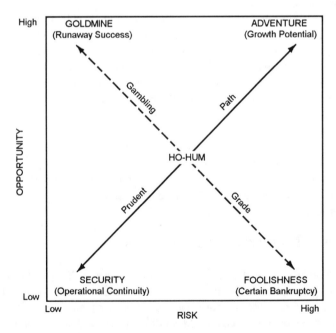

Figure 15-1. System strategy spectrum. (Reprinted with permission from V. Grose, *Managing Risk*. Englewood Cliffs, NJ: Prentice-Hall, 1987.)

tunities for traveling to earth orbit. While it was aware that such travel had attendant risks, NASA advertised the shuttle as a virtual "bus to space"—carrying commercial payloads, experiments from foreign nations, and even a senator and a congressman. NASA was progressing up the Prudent Path of Figure 15-1, demonstrating more and more opportunities for the shuttle while apparently keeping the associated risks under control.

After the *Challenger* was destroyed, the effect for NASA was to slide backward down the Prudent Path—toward lower risk and lower opportunity. The next shuttle flight could have been successfully launched almost immediately after the accident if the launch pad temperature had been a bit warmer. But the "subtle and unrealized" factors—along with the "obvious and recognized"—were forced out into the open by the Rogers Commission investigation. This public exposure forced NASA to retreat to a lower combination of risk and opportunity.

The common denominator to risk in modern society is complexity. As life becomes more complex, specialized information is required to understand, evaluate, and decide issues involving risk. Complexity also forces specialization—in education, business, religion, politics, technology, and law. The net effect of specialization is a dissected, fragmented view of life rather than an integrated one. Each specialty has its own language, culture, and values, even its own risks.

As any subject or field becomes divided and subdivided into narrower and narrower specialties, a counterforce inevitably arises to cope with the disintegration of unity brought on by specialization. This compensative force tends to be interdisciplinary or generalist in nature—trying to collect and embrace all the multiplying specialties. Because it will have to sacrifice in-depth understanding of each specialty in order to communicate with them, this counterforce of generalization will also introduce new risks.

Thus, complexity produces a diverging rather than a converging set of risks. The number and diversity of risks in our society appear to be ever-increasing—with no likelihood of reversing the trend. Figure 15-2 is a brief list of risks, most of which can be traced to complex technology in our sophisticated society. Of course, the list we all face every day is much longer.

MODERN SOCIETY'S RESPONSE TO RISK

The 1984 release of lethal methyl isocyanate gas in Bhopal, India, the *Challenger* explosion, and the Soviet nuclear catastrophe at Chernobyl are memorable examples of what some are calling technology's "dark side." A news media poll taken in May 1986 revealed that although most Americans still believe that the products of science and technology do more good than harm, the margin had dropped in three years from 83

SOCIETAL RISKS
Bernard L. Cohen
University of Pittsburgh

Risk	Days of Life Expectancy Lost
Being male rather than female	2700
Smoking one pack of cigarettes per day	2200
Remaining unmarried	1800
Working as a coal miner	1500
Overeating by 200 calories a day	400
Riding in automobiles (10,000 miles per year)	200
Being murdered	90
Not using seat belts	50
Driving small cars	50
Drowning	40
Being one pound overweight	30
Fire	30
Air pollution (all sources)	25
Being poisoned	20
Consuming electricity from burning coal	15
Choking on food	12
Being asphyxiated	7
Being struck by a falling object	6
Being struck by lightning	6
Being electrocuted	6
Consuming one diet drink per day	2.5
Being bitten by an animal or insect	0.3
Consuming nuclear energy	0.05

Figure 15-2. Societal risks. (Reprinted with permission from V. Grose, *Managing Risk.* Englewood Cliffs, NJ: Prentice-Hall, 1987.)

percent to 72 percent. And [the number of] those who think that twenty years from now technology will do more harm than good has risen to almost 25 percent.

This fear—known as *high-tech anxiety*—is triggered by a wide range of risks. Other sources include fears of nuclear war, runaway genetic engineering, toxic wastes in landfills, computer security in financial institutions, nuclear waste disposal, and commercial aviation safety. Government regulation of these potentialities always involves a compromise between promoting progress via new technology and protecting the public. Worst-case scenarios are thought to be too frightening for the average person, so they are often banned or known only by a select few.

In a classic and oft-quoted treatise published in *Science,* "Social Benefit versus Technological Risk," Chauncey Starr pointed out several interesting premises about risk in our society that can be empirically validated.

Starr first divided risks into two categories: voluntary and involuntary. For the voluntary risks, you subject yourself, as an individual and by your own decision, to a risk. Examples might be hunting, flying your own airplane, or skiing. Involuntary risks, on the other hand, are imposed on you by society without your consent. These could include military service, nuclear power generation, or highway bridge design. Starr's analysis revealed that *the average person will accept roughly 1,000 times as much risk voluntarily as he or she will involuntarily.* As Starr so aptly puts it, "We are loath to let others do unto us what we happily do to ourselves."

He validated this 1,000:1 ratio in a variety of ways. First, certain activities were measured in terms of their probability of fatalities per hour of exposure versus the average annual benefit a person would receive by that exposure. Typical activities included traveling by automobile, game hunting, or using electrical power. In all cases, the risk of voluntary activities exceeded that of involuntary ones by 1,000 times.

Another way this vast disparity between voluntary and involuntary exposure to risk was demonstrated was by comparing mining accident rates with hourly wages paid to miners. Starr showed that the risk had a 10^3 relationship—in other words, people can be paid to take voluntary risks, and the pay scale will closely follow a third-power relationship to any risk they would involuntarily accept.

Starr further showed that for most of us acceptability of risk is inversely related to the number of people who participate with us in that risk. Automobiles, for example, were originally considered to be sporting vehicles—driven by daredevils. But as more and more people began to drive, the amount of risk considered to be acceptable (in terms of fatalities per person-hour of exposure) was reduced. Moreover, tolerable risks for both general and commercial aviation demonstrate this same idea by continuing to drop as more of us utilize those modes for transportation.

A corollary to this relationship between acceptable risk and numbers of people involved is that advertising has a positive influence on the acceptance of greater risk by the general public. Whether or not misery loves company, there seems to be within us a subconscious willingness to take risks that we can see others also accepting. Without question, the superabundance—and cleverness—of alcoholic beverage advertising can be credited with the rising per capita consumption of alcohol, despite the alarming dangers of drunken driving, physical diseases, and sociopsychological ills it produces.

As a final glimpse of modern society's response to risk, Starr demonstrates that the probability of dying from disease seems to be, psychologically, a benchmark or yardstick in setting the limit of voluntarily accepted risk. In most sporting activities (whether skiing, race driving, hunting, or athletic contests), the risk of death is surprisingly close to the risk of death from disease. It is as though we all have a subconscious computer that sets our courage at a level equal to, but not exceeding, the statistical mortality due to involuntary exposure to disease. Could this unique probabilistic barrier be that fine line that separates boldness from foolhardiness?

ALTERNATIVES FOR RISK

On the tenth anniversary of the Surgeon General's declaration that tobacco smoking is harmful to health, a course on System Safety was being taught for George Washington University at the Navy Safety Center in Norfolk, Virginia. The twenty-five students, all mature adults, had become good friends while taking the course. About half of them were active smokers.

Though only the dim-witted could have thought smoking to be *beneficial* prior to the Surgeon General's pronouncement, open discussion erupted in the classroom as to whether smoking was actually *harmful*. "Perhaps he has overstepped the bounds of scientific propriety to make such a categorical statement," some argued.

After some friendly but testy disputing about the risks of smoking, someone suggested that it was a good focal point for resolving whether or not people ignore conclusive evidence about risks. Smokers were asked if they would tell the class *why* they smoked. They all agreed to do so.

As amusing as the responses were, they provided an excellent summary, as shown in Figure 15-3, of seven excuses people use to ignore conclusive evidence of risk. These excuses are not used only to defend smoking. The same answers can be heard for not using seat belts, installing smoke alarms in your home, keeping a list of your credit card numbers, diversifying your investments, getting someone else to drive after you have had a drink, or preparing a will. They are universal expedients for dismissing conclusive evidence of risk.

FORCE-FIELD ANALYSIS

Force-field analysis is a technique developed by Kurt Lewin for diagnosing situations in which people exist. Lewin assumes that in any situation there are

---

Procrastination: "I'll quit next week."
Social pressure: "Everyone else is doing it."
Head-in-sand: "I haven't decided yet."
Resignation: "I'm addicted, so I give up."
Expert example: "My doctor still does it."
Rationalization: "It's no worse than any other forms."
Reaction: "If people would quit bugging me, I'd quit."

Figure 15-3. Excuses for ignoring conclusive evidence. (Reprinted with permission from V. Grose, *Managing Risk.* Englewood Cliffs, NJ: Prentice-Hall, 1987.)

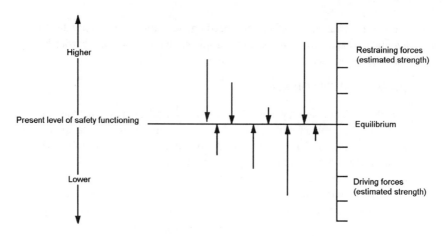

Figure 15-4. Driving and restraining forces in equilibrium.

both driving and restraining forces that influence any change that may occur. Those things that are pushing toward change are the driving forces. Those things that are preventing change are the restraining forces. Restraining forces to improved safety might be apathy, hostility, and unguarded machines. Driving forces might be an internal desire to work safely, the supervisor's honest wishes for the worker's personal safety, and so on. Equilibrium is reached when the sum of the driving forces equals the sum of the restraining forces, as shown in Figure 15-4. This equilibrium, or present level of safety activity, can be raised or lowered by changes in the relationship between driving and restraining forces (Figure 15-5).

An illustration of this in a productivity setting is provided by Hersey and Blanchard:[7]

> For illustrations, let us look again at the dilemma of the new manager who takes over a work group where productivity is high but whose predecessor drained the human resources (intervening variables). The former manager had upset the equilibrium by increasing the driving forces (that is, being autocratic and keeping continual pressure on subordinates) and thus achieving increases in output in the short run. By doing this, however, new restraining forces developed, such as increased hostility and antagonism, and at the time of the former manager's departure the restraining forces were beginning to increase and the results manifested themselves in turnover, absenteeism,

---

7. From P. Hersey and K. Blanchard, *Management of Organizational Behavior: Utilizing Human Resources*, 3rd edition. © 1977, pp. 124–125. Reprinted by permission of Prentice-Hall, Inc., Englewood Cliffs, NJ.

Perception of Risk

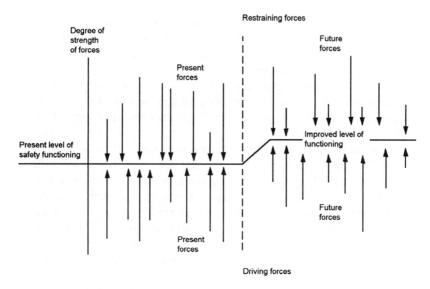

Figure 15-5. Force-field analysis.

and other restraining forces, which lowered productivity shortly after the new manager arrived. Now a new equilibrium at a significantly lower productivity is faced by the new manager.

Now just assume that our new manager decides not to increase the driving forces but to reduce the restraining forces. The manager may do this by taking time away from the usual production operation and engaging in problem solving and training and development. In the short run, output will tend to be lowered still further. However, if commitment to objectives and technical know-how of the group are increased in the long run, they may become new driving forces, and that, along with the elimination of the hostility and the apathy that were restraining forces, will now tend to move the balance to a higher level of output.

Managers are often in a position where they must consider not only output but also intervening variables, not only short-term but also long-term goals, and a framework that is useful in diagnosing these interrelationships is available through force field analysis.

Hersey and Blanchard also discuss the integration of goals and effectiveness, which seems particularly relevant to safety:[8]

---

8. Ibid.

The extent that individuals and groups perceive their own goals as being satisfied by the accomplishment of organizational goals is the degree of integration of goals. When organizational goals are shared by all, this is what McGregor calls a true "integration of goals."

To illustrate this concept we can divide an organization into two groups, management and subordinates. The respective goals of these two groups and the resultant attainment of the goals of the organization to which they belong are illustrated in Figure 15-6.

In this instance, the goals of management are somewhat compatible with the goals of the organization but are not exactly the same. On the other hand, the goals of the subordinates are almost at odds with those of the organization. The result of the interaction between the goals of management and the goals of subordinates is a compromise, and actual performance is a combination of both. It is at this approximate point that the degree of attainment of the goals of the organization can be pictured. This situation can be much worse where there is little accomplishment of organizational goals, as illustrated in Figure 15-7.

In this situation, there seems to be a general disregard for the welfare of the organization. Both managers and workers see their own goals conflicting with the organization's. Consequently, both morale and performance will tend to be low and organizational accomplishment will be negligible. In some cases, the organizational goals can be so opposed that no positive progress is obtained. The result often is substantial losses, or draining off of assets (see Figure 15-8). In fact, organizations are going out of business every day because of these very reasons.

Figure 15-6. Moderate divergence of the goals of management, the organization and subordinates: moderate degree of attainment. (From P. Hersey and K. Blanchard, *Management of Organizational Behavior: Utilizing Human Resources,* 3rd ed. Englewood Cliffs, NJ: Prentice-Hall, 1977.)

Perception of Risk   *241*

Figure 15-7. Extreme divergence of the goals of management, the organization, and subordinates: little degree of attainment. (From P. Hersey and K. Blanchard, *Management of Organizational Behavior: Utilizing Human Resources,* 3rd ed. Englewood Cliffs, NJ: Prentice-Hall, 1977.)

The hope in an organization is to create a climate in which one of two things occurs. The individuals in the organization (both managers and subordinates) either perceive their goals as being the same as the goals of the organization or, although different, see their own goals being satisfied as a direct result of working for the goals of the organization. Consequently, the closer we can get the individual's goals and objectives to the organization's goals, the greater will be the organizational performance, as illustrated in Figure 15-9.

Figure 15-8. Destructive divergence of the goals of management, the organization, and subordinates: loss of attainment. (From P. Hersey and K. Blanchard, *Management of Organizational Behavior: Utilizing Human Resources,* 3rd ed. Englewood Cliffs, NJ: Prentice Hall, 1977.)

Figure 15-9. Little divergence of the goals of management, the organization, and subordinates: high degree of attainment. (From P. Hersey and K. Blanchard, *Management of Organizational Behavior: Utilizing Human Resources,* 3rd ed. Englewood Cliffs, NJ: Prentice Hall, 1977.)

One of the ways in which effective leaders bridge the gap between the individual's and the organization's goals is by creating a loyalty to themselves among their followers. They do this by being an influential spokesperson for followers with higher management. These leaders have no difficulty in communicating organizational goals to followers and these people do not find it difficult to associate the acceptance of these goals with accomplishment of their own need satisfaction.

ACHIEVING CHANGE IN PEOPLE

If we are to change workers' perceptions about safety, we must try to change their attitudes and beliefs about what behaviors are appropriate in the work situation. We can do this in two ways: (1) by working one-on-one with them and (2) by influencing the group.

Influencing the Individual Worker

To use Kurt Lewin's force-field-analysis concept, we must understand the forces acting on the individual. Euninger has suggested the following approaches as a feasible way to do this (and they work well with any worker, male or female):[9]

1. *Emphasize and increase the advantages and satisfaction associated with working safely.* The idea is to strengthen a man's general readiness to choose the safe alternative in any situation by increasing the incentives to work safely (or by increasing his awareness of already existing incentives).

---

9. Reprinted by permission from M. Euninger, *Fundamentals of Accident Prevention for Supervisors,* Pittsburgh, PA, Normax Publications, 1962.

Compliment safe work. When you see a man doing something extra to work safely, compliment the man. Let him know you are aware he is making an effort to work safely.

Express appreciation for exceptional performance. Properly expressed appreciation adds further incentive for such men to work safely. Furthermore, the word gets around. Other men hear about such things. A climate is created for more willing cooperation.

Request commendation for deserving men. If you have a man who is deserving of special recognition because of his accident-free record, or his outstanding cooperation on safety matters, or something comparable, bring him to the attention of higher supervision.

Emphasize the personal gains of working safely. In your personal and group safety contacts, you have many opportunities to point out the personal gains of working safely. Do so frequently. In personal contacts, stress the gains that you think will impress the man. Stress the family angle for the family man. Stress the reputation angle with the proud man. Stress the promotion angle with the ambitious man. Tie working safely to what you know he wants.

Emphasize the job gains of safe methods. Whenever a recommended safe procedure has additional benefits, such as requiring less time, or requiring less exertion, or resulting in fewer quality defects, point out such additional incentives when you instruct the procedure [*sic*] to new men. Increase their awareness of the total benefits.

Always explain the why of a required safe practice. They are more likely to cooperate if they have been reasoned with than if they have been ordered. Besides a complete understanding has incentive value.

Encourage employee safety participation. Make it a practice to ask men for their opinions and suggestions concerning safety problems.

2. *Eliminate, where possible, any existing disadvantages or dissatisfactions associated with working safely.* The idea is to assure that situations will be sized up in favor of the safe choice because the deterrents against [*sic*] the safe choice have been eliminated.

Identify troublesome safe practices. Be alert for safe practices and procedures that are repeatedly ignored or violated. Try to accumulate a list of such practices and procedures.

Identify the deterrents. Select a troublesome practice for study. Determine why men are shying away from the safe practice.

Solve the deterrent problem. Study the troublesome safe practice for some way to minimize or eliminate the deterrents that are causing men to reject it. Discuss the problem with the men involved. Get their ideas on what can be done. Even if you get nowhere, you show them your interest in eliminating their problem.

Install the change. Solving the new problem in your mind or on paper is not installing the solution.

3. *Emphasize and increase the disadvantages and dissatisfactions associated with working unsafely.* The idea is to strengthen a man's readiness to reject the known unsafe alternative in any situation by increasing the deterrents against working unsafely (or by increasing his awareness of already existing deterrents).

    Stress the long run certainty of accidents. In your initial job instructions and in your regular safety contacts, stress the point that accidents have a habit of catching up with unsafe practices, given time.

    Stress severity potential where it is real. Men often fail to appreciate fully to potential injury severity associated with an unsafe practice. They need to be reminded of just how severe the possible injuries might be.

    Always correct observed unsafe practices. Most men don't like to be corrected for things done wrong in their work. That makes the possibility of a correction a useful deterrent. If you show your disapproval of unsafe practices by prompt on-the-spot correction, you create a climate of disapproval for such practices.

    Get key men to show their disapproval of unsafe practices. Expressions of disapproval coming from key men can be powerful deterrents.

    When necessary, show readiness to discipline. There is a limit to how often a man should be corrected by simple reinstruction for repetition of an unsafe practice. When that point is reached, a supervisor must take a firm stand. He must show that he is ready to initiate or request disciplinary action.

    Always explain why an unsafe practice is unsafe. Don't merely label a practice unsafe, and expect your label to be accepted at face value. Also, don't assume your men always know why a given practice is unsafe. Always explain why and how the practice is unsafe.

    Tell them about recent accidents. Make it a practice to keep your men informed about serious plant accidents when such accidents are related in any way to the work they do.

4. *Eliminate, where possible, any existing advantages or satisfactions associated with working unsafely.* The idea is to eliminate any elements in either the situation or the man that act as an incentive to work unsafely.

    Determine the underlying cause of the man's behavior. Try to figure out what the man gains from the unsafe practice. The nature of the unsafe practices will often provide the clue. Show-offs, for example, are easy to spot. If the unsafe practices don't tip you off, ask him. His answer will often provide the clue.

    Eliminate or nullify the underlying cause. To the extent that your situation permits, give a little extra attention to such a man. Try to eliminate his "need" to work unsafely to gain such questionable satisfactions. For example, the man who acts unsafely to show off wants attention. Try giving him some. Use the right occasions to discuss a job with him, or to compliment him for something he does well, or to get his opinion on something.

In addition to these thoughts from Euninger, one of the best ways to achieve change in the individual is through involvement; allow him or her to participate in the decision-making process. One of the classic studies in this area was done by Coch and French in an American factory. They found that when employees were allowed to discuss and participate in proposed technological changes, productivity increased, and resistance to the changes was decreased markedly.

The belief of most workers that an accident will not happen to them is normal and a real safety problem. It is a perception problem, and one that in some ways we try hard to instill in people as they are growing up. We must, if we are to deal with this problem, deal with workers' perceptions and attempt to change them to perceptions of hazards and of what their behavior must be to achieve their own personal safety. To do this we can work with individuals or with groups.

REFERENCES

Cartwright, D. Achieving change in people: some applications of group dynamics theory. *Human Relations* 4:381–392, 1954.

Coch and French. In P. Lawrence, How to deal with resistance to change. *Harvard Business Review* 32(3):49–57, 1954.

Euninger, M. *Fundamentals of Accident Prevention for Supervisors.* Pittsburgh, PA: Normax Publications, 1962.

Grose, V. *Managing Risk.* Englewood Cliffs, NJ: Prentice-Hall, 1987.

Hastorf, A. and H. Cantril. They saw a game: a case study. *Journal of Abnormal and Social Psychology* 49:129–134, 1954.

Hersey, P. and K. Blanchard. *Management of Organizational Behavior: Utilizing Human Resources,* 3rd ed. Englewood Cliffs, NJ: Prentice-Hall, 1977.

Katz, D. The functional approach to the study of attitudes. *Public Opinion Quarterly* 24:163–204, 1960.

Lewin, K. *Field Theory in Social Science,* edited by D. Cartwright. New York: Harper & Row, 1951.

Lippitt, G. *Visualizing Change.* La Jolla, CA: University Associates, 1976.

Mason, C. Manhood versus safety. In *Directions in Safety,* edited by D. Weaver and T. Ferry. Springfield, IL: C. C. Thomas, 1975.

Toch, H. and M. MacLean. Perception and communication: a transactional view. *Audio-Visual Communication Review* 10:55–77, 1967.

Chapter *16*

Reducing Decisions to Err

This type of human error reduction is perhaps the most difficult for managers to deal with. We can fix traps through engineering; we can do much to reduce overload through administrative controls; but it seems harder to most of us to work within the worker's head, to change attitudes, to alter the way another person thinks.

SETTING EXPECTATIONS

We influence behavior every day in nonsafety situations. If we want to change or increase how much a worker produces, we simply change a production quota; we set a new standard, a new expectation. Once that expectation is met, we reward for that performance; in short, we change behavior.

So one way we change the worker's decision to err is to do what we as managers do the best:

1. Define what is wanted (expectations—or standards).
2. Measure to ensure that those standards are met.
3. Reward for standard completion.

POSITIVE REINFORCEMENT

As shown in Figure 13-1, favorable consequences for desired performance build safe behavior. This simple approach is the most overlooked one in safety. In the perception survey (Figure 5-3), recognition for desired performance was the lowest response. In almost every perception survey we have done, recognition is the worst category. We usually do not recognize safe performance. If we do, we get great results in a short period of time.

A few psychologists have experimented with positive reinforcement in safety applications. Roger Brown, while at the University of Wisconsin, experimented with tokens as reinforcers and suggested implementing a practical program in industry. A program at the Injury Control Research Laboratory under the direction of Dr. Robert McKelvey showed interesting potential for the concept.

Another experiment by the Human Development Corporation showed tremendous reduction in occupational injuries using the reinforcement concept in the shipbuilding industry. Also reinforcement was used in safety training by Rubinsky and Smith at the University of Rhode Island and found to be more successful than traditional safety training.

A Project Study

In the spring of 1982, the Association of American Railroads, funded by the Federal Railroad Administration, embarked upon an experiment to learn if the concept of positive reinforcement would be feasible and effective in an industry as atypical as railroading.

A project study team was formed composed of safety directors of some of the major U.S. railroads, with me as an outside consultant. In earlier years, this project study team had done major work in administering a survey to several pilot railroads to better understand the perception of workers, supervisors, and three levels of managers of railroad safety programs.

Phase II of this research effort was to determine if some new approach might impact those accidents on the job caused by human error. Traditionally railroad statistics show that a major portion of all accidents are human-error-caused.

On each of two pilot railroads a group of between 75 and 150 first-line supervisors was selected to receive experimental training in human behavior, accident causation, human error reduction, positive reinforcement techniques, and employee assessment techniques. The cooperation of their superior was also sought to assure that the techniques taught would be used by the supervisors once they were back on the job after the training.

Measures of Performance

For each of the four divisions as many measures of safety performance were taken as possible, both before and after, to best judge the results. These measures were used:

Reducing Decisions to Err 249

1. Knowledge was attained by supervisors as a result of the training experience. Before and after tests were used.
2. Reactions to the training and to the technique of positive reinforcement were asked, initially and after three months of using the technique.
3. Performance of supervisors, before and after three months, was measured by asking each to report the number and kinds of activities engaged in the week prior to training, and the week prior to the three months follow-up training.
4. Performance of the work force as measured by safety sampling, a statistically valid sample of worker behavior, was sampled before training, after three months, and after six months. Samples were taken in all cases by safety professionals, and the same samplers were used throughout the entire study period to ensure accuracy and uniformity.

Accident results were not used as a measure of whether or not the program was effective. Whether or not accidents happened is simply an invalid measure of effectiveness because of the small-size group and the short length of the study. One division's record for six months is not a large enough sample to yield the facts needed to judge the effectiveness of any program.

Results Obtained

Figure 16-1 shows the results for these measures as written up by the project committee in the executive summary and as presented to the top executives of each railroad by the consultant. The primary measure used to determine program effectiveness was the safety sampling results. As shown in Figure 16-1, in the experimental division there was an improvement of performance of 40% in one railroad and 49% in the other. Changes in both control divisions were insignificant. All measures showed significant improvement in the experimental divisions compared to the control division counterparts.

As indicated in section V in Figure 16-1, the accident record, while not regarded as a judge of results, did improve considerably more in experimental than in control divisions. One railroad showed a 51% reduction in the accident record for the period of the study compared to the previous period.

It would seem that this training did in fact have an impact in both experimental divisions. The project study committee of the Association of American Railroads stated that the following statements can be made on the basis of the results obtained:

> It is possible to significantly improve safety performance through educating first-line supervisors to the human behavior problems associated with acci-

				RRI	RRII
I.	Knowledge—gain as measured by pre- and post test scores	A.	Average pretest score	4.54	3.58
		B.	Average post-test score (same items)	23.57	28.50
		C.	Gain	519%	769%
II.	Reaction to training and use of techniques taught	A.	Average of reaction to training session (scale 0 –5)	4.38	4.23
		B.	Positive reaction to being asked to use technique (scale 0 –5)	4.29	4.18
		C.	Positive reaction to using technique (3 mos. later 0 –5)	4.04	4.05
III.	Performance reported by supervisors	A.	Reported increase in use of techniques taught—as measured by what supervisors described themselves as doing before and after training	40%	14%

				% of Unsafe Behavior Observed			
				RRI	RRI Control	RRII	RRII Control
IV.	Performance of work force as measured by sampling of worker behavior before and after training supervisors		A. Prior to training	35%	34%	39%	31%
			B. Three months after training	20%	34%	23%	29%
			C. Six months after training	21%	31%	20%	30%
	Improvement in performance			40%	8%	49%	3%

V. Accident performance for the two areas was influenced by a number of variables not related to the study. A short-term comparison of reported accidents for the experimental and control groups on both properties shows both "experimental groups" experiencing fewer injuries than did their respective "control groups."

Figure 16-1. Summary of study results.

dents, training them to apply positive reinforcement techniques in their daily activities and providing support through the organization for their efforts.

Supervisors Required to Use Training

More important that this training, however, is the fact that supervisors were required to use some of the training given when back on the job. In each case the division manager was asked to devise a simple system of accountability where each supervisor had to report weekly to his boss his utilization of the concepts learned. In each case the supervisors were asked to observe workers as usual and to report to their boss the number of observations made each week.

Second, supervisors were asked to act following each observation. If the worker observed was working unsafely, the supervisor was asked to deal with the infraction as he normally would. If, however, the worker was working safely, the supervisor was required to act also; to contact the worker and positively reinforce that desired safe behavior—to do something with that worker that might make that worker more likely to work safely in the future.

Each week each supervisor reported to his boss on a small 3×5 card the number of observations made and the number of positive reinforcements given.

Conclusions

This study (as well as many similar ones) seems to show that we can significantly impact an individual's decision on whether or not to work unsafely (to err); and the keys seem to be training and a system to ensure that supervisors are in fact using their training—a system that requires them to observe, contact, and reinforce desired behavior.

Second, we recognize in such a study the power of the informal group; we assess where we are and attempt to build strong cohesive groups, as discussed in Chapter 13, with goals in tune with ours in management. This is usually done through employee participation and involvement that aim to establish a feeling of ownership in the workers.

With employee involvement, a number of techniques that can be used to assess and improve the system will bring improved behavior. Force field analysis (discussed in Chapter 15), statistical process control tools, and others are being used today with excellent results. Dr. Euninger's ideas in Chapter 15, although over 30 years old, are as relevant today as when they were first proposed.

The worker motivation model presented earlier (Figure 11-1) also shows many of the factors involved in a worker's decision to work safely or unsafely.

PART V | *Roles*

Chapter *17*

Human-Error-Reduction Concepts

This part of the book has to do with building a system to control and to reduce human error, whereas previous chapters have been academic—theoretical. There is no question that we can reduce the probability of human error. We know what causes it; thus we can control it. Once we understand the cause of almost anything, we can control it.

There are many causes of human error, most of which we can control. Although "to err is human," we can lessen the probability of error on the job in a variety of ways. So our discussion now turns to the lessening of the probability of human error: what we can do to lessen that probability in an industrial organization, and, more important, who is supposed to do what to make it happen.

The important thing to remember is that to err is human only in those situations where erring pays off. It is human to err when there is a payoff, but it is far less human to err if the payoff is totally negative and even painful. It is even less human to err if there is no payoff whatsoever. When we remove the rewards for unsafe behavior, that behavior will disappear. When we begin to reward safe behavior, that behavior will become common.

Though the above observation is stupidly simplistic, it also is completely true. The stupidity is not in its simplicity, but rather in our inability to see reality. People do those things for which they see a payoff. We are "paid off" for getting the product out, for serving the customer, for getting the job done on time. We are seldom (if ever) paid off for not getting injured each day.

The solution to human error control is simple, making it almost impossible for industrial management to accept it. The answer is: "Give safety performance a payoff." If safety is a big priority, then:

- Make executives pay attention to it.
- Make it a big goal of middle managers.
- Make it an integral part of the daily activities of lower managers.
- Allow workers a say in it.

- Build it into the performance appraisal systems.
- Build it into the daily numbers game.

The above ideas perhaps seem too general. They are the concepts we must believe in to build the organizational culture we discussed earlier. In addition to these concepts, it is necessary to have a system to force those actions that make the concepts real. In any organization just having the concepts is not enough; they are no more than the "vision," as managers like to call it today. Without a system to force action, the vision is no more than a pipe dream.

The system that forces action which will change a vision into a culture almost always starts with accountability (a system that defines tasks—who is to do what, when) and then measures to ensure task completion so that rewards can follow.

In this part of the book we will outline the system and define a number of tasks that can be performed at each level of the organization to lessen human error. The system used to make anything happen is the same one I have propounded in every book I have written on safety management since 1967.

What Makes It Happen?

The first and most important principle of modern safety management is: "The key to line safety performance is management procedures that fix accountability." Another way to state this is: "What gets measured and rewarded gets done."

Figure 17-1 shows a supervisory performance model (for safety or any other performance). The model suggests that whether or not your supervisors do anything to satisfy their safety responsibility is dependent upon:

1. Their knowing what to do (task definition).
2. Their knowing how to do it (training).
3. Their knowing that their boss is measuring whether or not they do it.
4. There being a reward for doing it.

What drives performance is the supervisors' perception of what the boss wants done, their perception of how the boss will measure them, and their perception of how they will be rewarded for that performance. To restate what the research shows, these questions dictate supervisory performance:

- What is the expected action?
- What is the expected reward?
- How are the two connected?
- What is the numbers game (how measured)?
- How will it affect me today and in the future?

Human-Error-Reduction Concepts 257

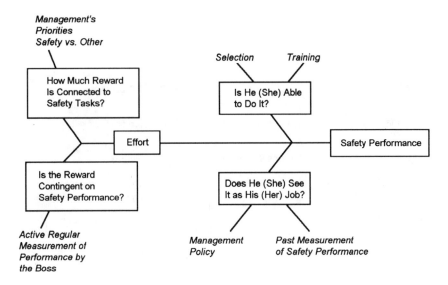

Figure 17-1. Supervisory safety performance model.

Any management system that defines, validly measures, and adequately rewards will work. Once a system is in place that ensures managerial performance, we can begin to focus on the tasks that people can do to lessen human error.

The human-error concept and the human-error-reduction (HER) model seem to suggest a number of possible controls that are available to us. Some of these are quite traditional, and others are somewhat new. Some seem usable right now, as we have the techniques needed to use them and, in some instances, have had them for years.

The purpose of this part of the book is to examine the controls that HER seems to lead to, and to discuss the kinds of controls that seem to be needed and the kinds of controls that seem to be feasible for use in our normal management systems.

Figure 17-2 lists the kinds of controls that seem to be indicated in the earlier chapters of this book. They are given in two categories, traditional and nontraditional. Traditional controls are those that we seem to be currently using, and nontraditional controls are those we are not currently using in safety programming.

TRADITIONAL CONTROLS

Many of the traditional controls are self-explanatory, in common use, well known to most safety professionals and students. Environmental analyses, usually made by industrial hygienists, are common in most safety programs. They

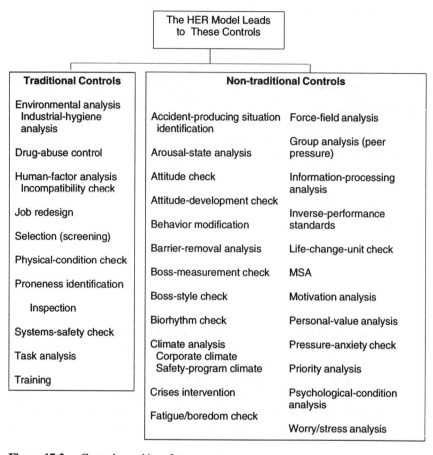

Figure 17-2. Controls used in safety programs.

usually deal with the load aspect of overload. We attempt, through environmental analysis, to ensure that physiologically the employees do not exceed the threshold limit value (TLV)—that they do not receive a load beyond their physiological capacity.

Drug-abuse-control programs are also common in many industries today. Their purpose is to reduce the probability of temporary reduced human capacity from drug and alcohol use on the job.

Human-factors controls are relatively well known to safety professionals, but are little used. While human-factors knowledge is of critical importance to safety people and safety is a primary reason for human-factors research, in actual industrial practice the two are not closely allied. There are few industrial organizations that employ human-factors specialists. Worse yet, few safety professionals have much of a working knowledge of human factors. In short, while human-factors controls fit into the traditional category, they really are not used much, if ever, in the day-to-day working life of the safety professional.

Worker participation is another traditional control that is grossly underused by safety professionals although some aspects of it have been utilized. Job safety analysis is used only to identify hazards, and then only for training purposes. Job redesign should be used to identify and remove accident-producing situations or to improve and enrich jobs by building additional meaning and worth into them. Safety is one area in which we can build job enrichment; employees who participate in safety decision making will not only reap the benefits of participative decision making, but will become infinitely better acquainted with the hazards of their jobs than any safety engineer ever could be.

Selection is one of our traditional safety controls although its value today is subject to some discussion. Employers' selection prerogatives are more limited than ever. Even if we could legally screen and discriminate effectively (the essence of selection), we have no way of determining which applicant will be a safe worker and which will be unsafe. Selection is limited in its effects on safety.

The use of physical examinations in the selection process is also a traditional control in safety programs, but such examinations are utilized less today than they were 30 years ago because of a number of laws limiting their use.

The identification of accident-prone personnel has been attempted traditionally by safety professionals, with almost no success. As indicated earlier in this book, it is a complex and confusing area. What is clear, however, is that if there is such a thing as accident proneness, at this point we have no real way of identifying who is accident-prone and who is not. For now, this control has limited value.

There is nothing more traditional and more used by safety professionals than inspection. It was our only approach to accident reduction initially, and is still perhaps our primary approach. Information on the subject abounds in the safety literature.

Systems safety is much like human factors. The field of systems safety is filled with concepts and techniques. Its use by the average industrial safety professional, however, is most limited. Government and large industries use systems safety, but almost no one else does. Apparently it is too complex and expensive to be used in typical safety programs in small industries. Or perhaps we simply have not learned the techniques that would make it more usable and applicable to the vast majority of safety programs.

Task analysis had been known as job safety analysis. With the new ergonomic thrust, it will be broadened considerably under the new title.

Training is as traditional as inspection. It is used everywhere and for almost everything. It is used when it should be for problems caused by lack of knowledge and skill, and it is used when it should not be for problems not so caused. Along with inspection, training is used almost to the total exclusion of any other control available to the safety professional.

Nontraditional Controls

Since so much information is available about all of the traditional controls, we will not discuss them here. Our emphasis will be on the nontraditional controls, those possible approaches to the reduction of accidents that the HER model seems to lead to. The categories of nontraditional controls are:

1. Techniques developed and used, but not traditionally used in safety.
2. Techniques discussed or researched in safety but not in use because we have not figured out how to use them.
3. Techniques discussed in nonsafety areas that will have applications in safety in the future.
4. New techniques.

Techniques Developed and Used, but Not Traditionally Used in Safety

In this category, two techniques might be considered: behavior modification and crisis intervention. These techniques are enjoying considerable success in fields totally separate from safety.

Behavior Modification

Behavior modification is being used in industry as well as many other areas to shape human behavior. Since human error is a type of human behavior, behavior modification is one of the ways we can change human behavior and reduce human error. Behavior modification might be considered an alternative to training, which, as previously indicated, is one of the safety professional's primary approaches. Although some safety textbooks have dealt with behavior modification in some detail, the safety literature has discussed it, and some institutions have taught it to safety professionals, it is seldom used by safety professionals. The author believes that behavior modification is a proven, solid technique that the safety professional might well put to use in certain situations to reduce the probability of human error.

The uses of this concept were discussed briefly in Chapter 16. Only one short experiment was shown there. However, the technique has been used a great deal since the first edition of this book. Table 17-1 shows the result of a number of studies.

Crisis Intervention

Crisis intervention techniques are being used today in nonsafety situations; this is a proven approach. At certain times workers are more predisposed to injury. As they are coping with other, perhaps personal, crises, they are subject to injury on

Table 17-1. Safe behavior reinforcement studies and results

Researcher	Year	Type Reward	Measures Used	Results Obtained	Percent Improvement
1. Komaki, Barwick & Scott	1978	Praise & Feedback 3–4x/wk	% Safe Behaviors	70% up to 95.8% 77.6% up to 99.3%	37% 29%
2. Komaki, Heinzman & Lawson	1980	Feedback Daily	% Safe Behaviors	34.4% up to 68.4% 70.8% to 92.3%	98% 30%
3. Krause	1984	Feedback	Unsafe Behaviors Lost Time Accidents Severity Rates		80% 39.3% 39.2%
4. Sulzer-Azaroff & DeSantamaria	1980		Accident Freq. Accident Freq. Hazards Hazards	15 to 0 45 to 33	100% 27% 29% 88%
5. Zohar	1980	Tokens for future use	% Use of Ear Plugs % Use of Ear Plugs % Use of Ear Plugs	35% up to 85% 35% up to 90% 50% up to 90%	143% 157% 80%
6. Sulzer-Azaroff	1978	Trading Stamps		Reduction (No more information)	
7. Fox	1976	Trading Stamps		Reduction (No more information)	
8. Calkins	1971	Cash Bonus & Paid Time Off	Vehicle Accidents Personal Injuries Dollars Saved	$5766 Saved—$1899 Spent	48.7% 41.0%
9. Hoppe & Terry	1983	Tangible Incentives	% Seat Belts Used % Seat Belts Used	36% up to 60% 60% up to 70.3% 65% up to 70%	67% 8% 8%
10. Rummier		Praise & Recognition	Lifting Practices	Improved (No more information)	
11. Smith, Anger & Uslan		Praise	Accident Rate	Improved in 5 Crews	
12. Felliner & Sulzer-Azaroff	1984	Praise & Feedback	% Safe Behaviors	78% up to 86% 79% up to 85%	10% 8%
13. Harrell		Self Monitoring	Number of Accidents	Reduced for 8 Mos.; Reversal Phase—Returned to Previous	
14. Komacki	1979	Feedback & Off Time			
15. Sundstrom	1984	Piece Rate to Monthly Rate	Accident Frequency		29%
16. McKelvey	1973	Financial Incentive	Number of Accidents		80%
17. Petersen	1983	Praise & Feedback	Safety Sampling	Two Railroad Divisions for Six Mos. with Control Groups	40% 49%

Table 17-1. *(Continued)*

Researcher	Year	Type Reward	Measures Used	Results Obtained	Percent Improvement
18. Rhoton	1980	Praise & Feedback	Violation	1–4 per mo. to 0	100%
19. Fox, Hopkins & Anger		Tokens	Accident Frequency		90%
20. Hopkins, Conrad & Smith		Feedback & Money	% Safe Behaviors	60% up to 100%	67%
			Housekeeping Rate	20 to 90	350%
			Housekeeping Rate	45 to 100	122%
21. Chokar & Wallin	1984	Feedback	% Safe Behavior	65% up to 81%	20%
				81% up to 95%	17%
22. Uslan & Adelman	1977	Praise	Injury Frequency		50%
23. Earnest	On-going	Teaching Behavior Management	Incidence Rate	4.0 down to 2.5 (Now Much Lower)	38%
			Dollars Saved	$1.5 million (Now More)	
24. Nasanen and Saari	1987	Visual & Verbal Feedback	Housekeeping Index # Accidents	62% to 75% FB to Foremen only	21%
				62% to 88% FB to Workers Also	42%
25. Krause, et al. Mfg.	1980	Visual Feedback	Unsafe Behavior		60 to 80%
Oil	1982	Visual Feedback	Safe Behavior	79% to 94%	19%
Plastic	1984	Visual Feedback	Safe Behavior	79% to 91%	15%
Chemical	1984	Visual Feedback	Safe Behavior	39% to 81%	108%
Utility	1985	Visual Feedback	Acc. Freq.		26%
Utility	1985	Visual Feedback	Back Injuries		67%
Chemical	1985	Visual Feedback	Safe Behavior		35%
Utility	1985	Visual Feedback	Safe Behavior		25 to 45%
Utility	1986	Visual Feedback	Acc. Freq.		42%
Trans.	1984		Acc. Freq.		66%
26. Petersen	1987	Verbal Recognition	Unsafe Behaviors	Eight Railroad Divisions Over 6 Mos. with Control Groups	11% to 28%

the job. Crisis intervention might be one way to help them avoid injury. A system in which someone intervenes prior to a possible accident is a feasible safety approach. Crisis intervention is used regularly in nonsafety situations; it just might be a major help in the safety program.

Techniques Researched but Not Yet in Use

We next look at some concepts and ideas that have not yet been tried in safety because they are not yet fully developed. Examples might be the identification of attitudinal state, the interpretation of biorhythmic cycle points, dealing with fatigue and boredom, spotting and dealing with an overload of the information-processing capacity, and using our knowledge of life-change units (LCUs).

The Attitudinal State

One aspect of overload is the attitudinal state in which the overload takes place. This is difficult to deal with on a practical basis. We know little about measuring attitudinal state, much less about how to change or develop attitudes.

Robert Mager's book *Developing an Attitude Toward Learning* contains a great deal of insight into attitudinal development that we can apply directly to the development of safety attitudes. Mager reported on a study he made several years earlier that was designed to determine students' attitudes toward different academic subjects and how they were formed. He found that a favorite subject area tends to become favorite because the person seems to do well at it; because the subject is associated with liked or admired friends, relatives, or instructors; and because the person is relatively comfortable dealing with the subject. Conversely, a least-favorite subject seems to become so because the person has a low aptitude for it, because it is associated with disliked individuals, and because the subject matter is associated with unpleasant conditions. We can see from these findings that an attitude toward a subject is molded by:

> The conditions that surround the subject.
> The consequences of coming into contact with the subject.
> The way that others react toward the subject (modeling).

Conditions. When a person is in the presence of subject matter, he or she should simultaneously be experiencing positive, not negative, conditions. This fosters positive feelings about the subject. If a neutral subject is presented to someone on several occasions while he or she is experiencing an unpleasant condition, that subject may become a signal that triggers an avoidance response. If a person is presented with a neutral subject and at the same time is enjoying pleasant conditions, that subject may become a signal for an approach response. For instance, Mager cites the familiar reaction of most people to a hypodermic needle.

Once we have experienced pain in the presence of the needle, it becomes a signal for an avoidance response; the mere sight of a hypodermic needle becomes a condition to be avoided.

How can we use this information in our safety programming? First of all, if we do not already know what our employees consider to be positive and negative conditions, we certainly ought to find out. For example, any form of punishment is obviously negative, most forms of social interaction are positive, competitive and game-type situations are positive, participation is positive, and being told what to do is negative. We should arrange our safety instruction so that while the employee is being instructed, he or she is experiencing as many positive conditions and as few negative conditions as possible.

Consequences. Whenever contact with subject matter is followed by positive consequences, the subject will tend to stimulate *approach* responses. Conversely, whenever contact with a subject is followed by negative consequences, the subject may tend to stimulate *avoidance* responses. This statement can be found in any freshman psychology text. It has been documented experimentally and practically, perhaps more than any other psychological principle.

Mager translates the principle into a classroom situation by asking us to picture ourselves as students in a situation where, when we correctly answer a question posed by the instructor, he or she smiles and says something like, "Good." When we answer the question incorrectly, the instructor makes a comment such as, "Well, let's look at the question again." In this situation, Mager suggests we would no doubt be willing to answer questions and come into contact with the subject matter again. In any case, this kind of interaction would not adversely influence our response toward the subject. Conversely, suppose each time we answer a question the instructor says, "Well, I see old Dumbo is at it again." How long would it be before we stopped raising our hand? How long do you think it would be before we began to think of excuses for not attending class?

When experience with a subject is followed by a positive (pleasant) consequence, the probability is increased that the subject will be approached again in the future. When experience with a subject is followed by negative (unpleasant) consequences, the probability is reduced that the subject will be approached in the future. If we want to increase the probability that a response will be repeated, we have to follow it up immediately with a positive consequence. If we want to reduce the probability that the behavior will occur again, we have to follow it up immediately with a negative consequence.

This all seems obvious, but perhaps one point is a little confusing. In one case, we seem to say punishment is bad; in the next breath, we say that application of negative consequences is good. Which is it? By negative consequences, we mean conditions. Punishment is not recommended. A safety program built on punishment will not succeed. It will create aversion to the subject of safety. Use of punishment puts safety and the safety professional in a bad light. Also, with a program based on punishment, safety rules must be tightly set to make the punishment fair. This is negative because it is "telling them what to do." Punishment following an unsafe act might work for the individual alone, but here the problem is whether we can be sure that he or she will no longer perform the act and not merely make sure we do not catch him or her at it the next time.

Positives and Negatives. What in a work situation is negative, and what is positive? Although it is not always possible to know whether an event is perceived as positive or negative for a given individual, some conditions and consequences are universal enough to provide us with some direction.

Mager suggests we define an "aversive" as any condition or consequence that causes a person to feel smaller or makes his or her work "dimmer." Here are some common causes of aversion (negatives) that might apply to safety and safety training:

1. Pain. Pain experienced on the job facilitates learning (the hard way).
2. Fear and anxiety. Fear and anxiety are caused by occurrences that threaten various forms of unpleasantness.
3. Frustration creators.
4. Humiliation and embarrassment.
5. Boredom.
6. Physical discomfort.

We can create positive conditions or consequences by:

1. Acknowledging responses, whether correct or incorrect, as attempts to learn, and following them with accepting rather than rejecting comments. ("No, you'll have to try again," rather than "How could you make such a stupid error?")
2. Reinforcing or rewarding subject-approach responses.
3. Providing instruction in increments that will allow success most of the time.
4. Eliciting learning responses in private rather than in public.
5. Providing enough signposts so that the worker always knows where he or she is and where he or she is expected to go.
6. Providing the worker with statements of our instructional objectives that can be understood right away.
7. Detecting what the worker already knows and dropping that from his or her training. (This prevents boredom.)
8. Providing feedback to the worker's response that is immediate and specific.
9. Giving the worker some choice in selecting and sequencing subject matter, thus making positive involvement possible.

10. Providing the worker with some control over the length of the instructional session.
11. Relating new information to old information that is within the experience of the student.
12. Treating the worker as a person rather than as a number.
13. Using active rather than passive words during presentations.
14. Making use of those variables known to be successful in attracting and holding people's attention: motion, color, contrast, variety, and personal reference.
15. Allowing to teach only those instructors who like and are enthusiastic about their subjects (and workers).
16. Making sure the worker can perform with ease, not just barely perform, so that confidence can be developed.
17. Expressing genuine delight at seeing the worker.
18. Expressing genuine delight at seeing the worker succeed.
19. Providing instructional tasks that are relevant to our objectives.
20. Using only those test items relevant to our objectives.
21. Allowing workers to move about as freely as their physiology and their curiosity demand.

Modeling. Another way in which behavior is strongly influenced is through modeling (learning by imitation). The research on modeling tells us that if we want to maximize approach (rather than avoidance) tendencies in workers, we must exhibit that behavior ourselves. In other words, we must behave the way we want our employees to behave. When we teach one thing and model something else, the teaching is less effective than if we practice what we preach. The message for safety management is obvious.

Attitude Development. Let us summarize what is known about developing attitudes generally and about adapting that knowledge to safety:

1. There are three ways in which attitude (and behavior) toward a subject may be influenced: by the conditions associated with the subject matter, by the consequences of contact with the subject matter, and by modeling.
2. Exhortation, a procedure used regularly for safety, has seldom been very successful in influencing behavior.
3. Approach and avoidance behavior are influenced by the things we do and by the things we say.

Biorhythms

As indicated earlier, we are not sure whether biorhythms are relevant to industrial safety. There is much conflicting information. If biorhythms are relevant, we do not know yet how to apply the concept in safety programs. There are reports from companies here and in Japan about using biorhythmic information to schedule bus drivers and airline pilots. These reports indicate a lessening of accidents as a result, but there appear to be no controls to determine whether the reduced number of accidents is the result of this factor or of some other, unknown, factor. It seems that the reduction might simply be due to the Hawthorne effect (the fact of being experimented with). Knowledge of employees' biorhythms might, however, be useful in accident reduction.

Fatigue and Boredom Control

There is considerable information in the literature on the relation of fatigue and boredom to accidents; we are relatively sure that such a relationship does exist. Argyris proposed boredom and fatigue as accident causes and as due to the innate conflict between the needs of mature human beings and organizational characteristics. Even though we tend to believe in the relationship between accidents and these factors, we have yet to study this relationship conclusively or to develop any meaningful safety controls that relate to it. However, while we still do not have all the information we need on this, the 1980s and 1990s have provided us with a mass of anecdotal evidence. As companies have gone through the almost universal process of downsizing, or reduction in force (RIF), we have seen a severe increase in our accident frequency rates, probably related to the fact that most companies have required more work from fewer workers, through overtime and other means.

Information-Processing Overload

In the human-factors literature there are some studies on information-processing overload, but the concept is almost totally overlooked by safety professionals. Perhaps we are a long way from this kind of thinking except in crucial or critical human-performance areas (piloting, etc.). This seems, however, an area where safety professionals need to put more time and effort. A large percentage (the largest) of on-the-job serious injuries and fatalities occur on the highway, and many of these might well be due to information-processing overload at the time of the accident.

Life-Change Units

The concept of life-change units (LCUs) is known to be valid; unlike the biorhythm concept, this one is documented with over 25 years of scientific study. The relationship between personal stress and illness (and accidents) does exist. However, we do not have any techniques to deal with LCUs. It is interesting that companies throughout the world spend time and effort on a fuzzy concept like biorhythms, while a solid concept like LCUs lies virtually untouched. Certainly LCUs could be used in some way in work scheduling. More feasible would be to use the concept to provide lower-level managers with information to aid them in their day-to-day supervision of employees.

Techniques Discussed in Nonsafety Areas

In this category, a number of concepts offer some very rich possibilities for the safety professional. For example, we might build techniques related to (1) the concept of arousal state, (2) attitude development, (3) managers as barrier removers, (4) boss measures, (5) boss styles, (6) corporate climate, (7) analysis of employee force fields, and (8) group pressure.

Arousal-State Control

This area is important to safety, as we know that an individual can sustain considerable overload without incident if sufficiently aroused. Yet we do not know how to use this knowledge other than to employ those simplistic, traditional approaches that create and maintain the worker's interest in safety. An increased arousal state means fewer human errors, but we know of no organizations that are developing a technique to ensure the heightened arousal state of employees. This area has great potential for future experimentation in industry.

Attitude-Development Control

As indicated earlier in this chapter, we know quite well how attitudes are developed—through the conditions surrounding safety, through the consequences of coming into contact with safety, and through modeling of management. We know that the attitude each employee has toward safety (or anything else) is a function of the negatives or positives that the employee associates with it.

Knowing this, we should be able to determine what attitudes will be held by analyzing the negatives and positives associated with safety performance, safety training, and so on. We tend to overlook this kind of analysis in our normal, day-to-day operations. It might well be worth our time to step back and observe several employees as they go through orientation or as they sit in a safety meeting, and to use a checklist of negative and positive items to find out exactly what workers are experiencing as they are confronted with the subject of safety in our organizations.

Barrier Removal

Douglas McGregor described the Theory Y manager and the Theory X manager. The Theory X manager perceives his or her role as coercive: employees must be forced to work. The Theory Y manager believes that employees like to work, and that the role of the manager is to remove the barriers that lie in their way. If we agree with the Theory Y manager, perhaps we should spend our time appraising lower-level managers to see how well they are removing barriers to safe on-the-job behavior.

Boss Evaluation

Research in the behavioral sciences tells us that one factor in employee behavior, and the largest determinant of supervisory performance, is the way the person is measured by his or her boss. If this is true, how well are bosses in our organizations enforcing and evaluating safety behavior? The traditional measures used by safety professionals to determine whether or not a lower-level manager is performing (whether or not accidents have happened) are inadequate, both for determining past performance and for building better future performance. Perhaps the primary task for the safety professional is to determine better ways of evaluating managerial safety at all levels.

Boss-Style Analysis

There are all types of bosses—some with Theory X assumptions and some with Theory Y assumptions. There are, in managerial-grid jargon, 1,9 managers, 9,1 managers, 1,1 and 9,9 managers, and so on. Each has a different style, different beliefs, and different approaches. We ought to understand these types and be able to evaluate and analyze the safety leadership in our organizations.

Climate Analysis

As was discussed earlier, organizational and safety-program climates greatly affect the success of safety attempts. This is ongoing research to determine the relationship of the organizational climate to safety success. It appears that climate may well be the number one determinant of safety success. If so, it is imperative that we assess climate and determine how it will influence the safety program.

Force-Field Analysis

In Chapter 15 we discussed the concept of force fields, of analyzing the motivational field that employees are in to determine what they will do on the job.

This concept provides us with a way to figure out ahead of time what determines employee behavior.

Peer-Pressure Analysis

Research tells us the primary determinant of employee behavior is the informal group, or peer group. If we know the employee's group norms and standards, we can predict the employee's behavior.

New Techniques

The HER model leads to concepts not used before, to the development of new techniques not yet tried in safety. Some of these new techniques are managerial priority analysis, inverse performance standards, leveling analysis, and worker-safety analysis.

Managerial Priority Analysis

An important determinant of employee behavior and of supervisory performance in safety is where safety is placed in management's list of priorities. How does safety rate in comparison with other managerial goals such as production and quality? It is important to determine quite precisely the priority that safety has in reality, not just as expressed officially.

Inverse Performance Standards

High-level managers use performance standards to appraise the performance of subordinate managers, yet actual corporate performance is much more determined by the actions and performance of the top-level managers than of the subordinate managers. A valuable addition to this system would be an inverse performance appraisal system, whereby top-level managers are appraised by subordinate managers or by employees. This technique, which was new in our first edition, is now used in many organizations.

Leveling Analysis

Leveling means the replacement of the traditional superior–subordinate relationship with a more egalitarian relationship—an adult–adult relationship. Since the interpersonal relationship between manager and employee is an important determinant of the employee's behavior, it should be examined with respect to leveling.

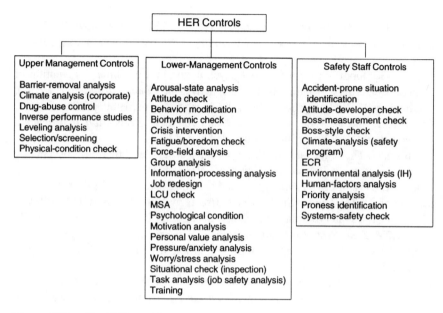

Figure 17-3. The HER model: control responsibilities.

Worker-Safety Analysis

We are familiar with job safety analyses, where we study the job to determine any hazards, or any accident-prone situations. Why not a worker-safety analysis, where supervisors study each worker and his or her current susceptibility to accident? We might include many concepts here: force-field analysis, LCU analysis, and perhaps biorhythmic analysis.

Summary

The HER model leads into many potential areas for accident control, most of them nontraditional. If we were to examine the controls indicated by the HER model and categorize them according to who would carry them out—upper management, lower management, or the safety staff—the result might look like the diagram shown in Figure 17-3.

REFERENCES

Aguilera, D., J. Messick, and M. Farrell. *Crisis Intervention: Theory and Methodology.* St. Louis: Mosby, 1970, p. 16.

Argyris, C. *Personality and Organization.* New York: Harper & Row, 1957.
Caplan, G. *Principles of Preventive Psychiatry.* New York: Basic Books, 1964, p. 38.
Caplan, G. Opportunities for school psychologist in the primary prevention of mental disorders in children. In *Perspectives in Community Mental Health,* edited by A. Bindman and A. Spiegel. Chicago: Aldine, 1969, p. 421.
Faucett, H. and S. Wood, eds. *Safety and Accident Prevention in Chemical Operations.* New York: Interscience, 1965, p. 47.
Jacobson, G. The scope and practice of an early-access brief treatment psychiatric center. *American Journal of Psychiatry* 121(12):1180, 1965.
Mager, R. *Developing an Attitude toward Learning.* Belmont, CA: Fearon, 1968.
McGregor, D. *The Human Side of Enterprise.* New York: McGraw-Hill, 1960.
Mims, F. Crises intervention in accident prevention. *ASSE Journal*, August 1972.
Morley, W. Treatment of the patient in crisis. *Western Medicine* 77(3):82, 1965.
National Safety Council Publication. The problem employee. *National Safety News,* January 1970, p. 47.
Petersen, D. *Safe Behavior Reinforcement,* New York: Aloray, Inc., 1989.
Smith, R. Rebutting behaviorism, in *Industrial Safety and Hygiene News,* March 1995.

Chapter *18*

Line Management Roles

UPPER MANAGEMENT

We begin looking at human-error controls by looking at what upper-level management can and should do. The influences exerted by upper management are described as follows by Charles Handy:[1]

> By the time a man enters management, many businessmen believe it is too late to change his "character." To a great extent this belief is true; the experiences of childhood and adolescence are indeed crucial. Many of the troublesome attitudes and actions of managers which are typically blamed on "character," however, can be traced to management itself. In other words, although "character" is relatively enduring, many of management's "people problems" are partly products of its own making. Again and again, in my observation of industry, I find that the undesirable behavior of subordinates is precipitated or aggravated by the unintentional actions of their superiors. Such a statement is not news to subordinates—to them it is by now a cliché—but it is often disdained by their bosses as sheer rationalization.
>
> In one sense, this observation is a discouraging commentary on the practice of management today. In another sense, it is a reason to be encouraged. For, to the extent that "people problems" are created by management, it has the immediate power to solve them by changing its approach. It does not have to defer the solution until long-range training programs and education have had a chance to work.

---

1. Reprinted with permission from C. Handy, *Understanding Organizations,* Middlesex, Eng., Penguin, 1976.

Upper-level management creates "people problems" and has the power to control them.

How effectively do organizations deal with their people, their human resources? The questions that follow were written to help organizations evaluate their management of human resources:

1. How much money was spent last year to recruit and select people?
2. Was this expenditure worth the cost?
3. Does your organization have data on standard costs of recruitment, selection, and placement which are needed to prepare manpower budgets and to control personnel costs?
4. Was the actual cost incurred last year less than, equal to, or greater than standard personnel acquisition and placement costs?
5. How much money was spent last year to train and develop people?
6. What was the return on your investment in training and development?
7. How does this return compare with alternative investment opportunities?
8. How much human capital was lost last year as a result of turnover?
9. How much does it cost to replace a key person?
10. What is the opportunity cost of losing young, high-potential managers, accountants, engineers, etc.?
11. What is the total value of your company's human assets?
12. Is it appreciating, remaining constant, or being depleted?
13. Does your company really (I mean *really*) reward managers for increasing the value of their subordinates to the firm?
14. Do your compensation and other motivation-reward systems reflect an individual's present value to the firm?
15. Does your organization consider its investment in human resources when evaluating capital budgeting proposals requiring the allocation of people?
16. Does your firm assess the effects of corporate strategies upon its human resources in *quantitative* terms?

Before we examine the change process necessary for the implementation of some human-error controls, let us look at what Harry Levinson has to say about the causes for human error:[2]

> In this article I shall discuss six common management actions which lead to troublesome or problematic behavior among subordinates, and suggest supplemental or alternative actions which might be taken to avoid the difficulty.

---

2. Reprinted with permission of *Harvard Business News*. Excerpts from "Who Is to Blame for Maladaptive Managers?" by Harry Levinson (November–December 1965). Copyright © 1965 by the President and Fellows of Harvard College. All rights reserved.

This analysis is based on an examination of 287 cases presented by participants in 15 executive seminars held at The Menninger Foundation during the past nine years.

CASES ANALYZED

There were 287 cases available for analysis for this article. Each of these cases was submitted in written form, in advance of a week-long seminar, by an executive participating in the seminar, for purposes of discussion in a small seven-man group. The participant was instructed that his case should be about a problem between people, preferably one in which he was one of the parties, and one which had been particularly troublesome. He was asked to describe in the case the circumstances of the problem and the nature of the discussions between himself and the other person. He was free to write the case in any way he chose, disguising it sufficiently so that it was not identifiable. In the small group discussions he did not have to identify his case unless he wished to do so, although in such a small group it was often apparent which case belonged to which participant. I reviewed each of these cases and, on the basis of the most prominent aspect of behavior described, assigned it to one of five categories:

1. *Nonmanagement Problems*—Of the total submitted, 27 cases did not constitute individual problems; 25 more were not problems of executives (middle management and above, but not first-level foremen). These cases have not been used in this article.
2. *Mental illness*—This category includes 15 cases in which psychiatric illness had already been diagnosed or in which the symptoms as described by the presenter were so severe that they were clearly indicative of illness. In 4 of these cases the subjects were manipulative to the point of dishonesty, exploiting customers as well as subordinates, colleagues, and superiors, and clinically would be called pathological character disorders. These cases, too, are excluded.
3. *Hostility*—This is the most conspicuous symptom or underlying feeling in 125 of the remaining 220 cases. Of these cases, 15 cover a wide range of hostile behavior; the other 110 cases fall into four subgroups: (a) the authoritarian ones, (b) the angry ones, (c) the self-centered ones, and (d) the wounded ones.
4. *Limited Personalities*—This group includes 48 men classified into four subgroups: (a) the anxious ones, (b) the rigid ones, (c) the dependent ones, and (d) the impulsive ones.
5. *Misplacement*—In this group are 47 men who had either been placed in the wrong job (the hapless ones) or who had been outgrown by the job (the helpless ones).

> The categories were derived from an inspection of the cases; they had not been previously established. They reflect my interpretation of the behavior of the person described in the case. The categories, therefore, are based on the problem as the presenter saw it—on the data as he perceived and reported them, though not necessarily as he interpreted his own data. They are not diagnostic categories in the clinical sense. Nor are they mutually exclusive, for they have been evolved from that aspect of the individual's behavior which I understood from the case description to be the most salient.
>
> These cases, then, are reports by senior executives of problems they have had with subordinate executives. The acute or painful aspect of the problem as presented is the behavior of the subordinate executive. But as the discussion elaborates, it turns out that the senior executive also has a part in each of these problems. His part results from lack of self-scrutiny, indifference, inadequate understanding, or some other nonfunctional behavior. This is not to say that the senior executive or the organization "caused" the problem. There are always multiple causes for all behavior. Yet in each one of these groups of cases, senior executives have made certain repetitive managerial errors that relate specifically to the kind of behavior which the subordinate displays. This correlation between managerial error and subordinate behavior enables us to focus on the former as a means of pointing toward more proficient managerial practice.

Levinson also identified six management errors that create problems for subordinates:[3]

> The actions that I shall describe frequently seem to make sense at the time they are taken. But while they may be defensible in the short run, in the long run they turn into liabilities. From the standpoint of sound organizational growth, they must be judged as errors.
>
> Error #1: Encouragement of Power Seeking
> No single kind of subordinate pleases his superiors more than the man who is able to assume responsibility for a crisis task, jump to his task with zest, and accomplish it successfully with dispatch. Such men become the "jets" of industry, the "comers," the "shining lights."
>
> Naturally, higher management rewards such men for their capacity to organize, drive, and get results. Management therefore encourages them in their wide-ranging pursuit of personal power. The men are described as problems because they reportedly overdominate their staffs; they are unable to coach and develop subordinates; they concentrate decision making in their

---

3. Ibid.

own hands while driving their subordinates unnecessarily hard. In short, they are *authoritarian*. Their individual achievements have led to promotion, and the aura of their record has obscured for a considerable time the fact that they are now destroying or failing to build some part of the organization.

Error #2: Failure to Exercise Controls

Senior executives, according to some of our cases, often seem to condone behavior which is beyond the bounds of common courtesy. The result is devastating to those who are subject to such behavior and detrimental to the organization.

Error #3: Stimulation of Rivalry

Rivalry, by definition, is the essence of competitive enterprise. But in such an enterprise, where the desirable end product is the result produced by the organization, all effort should be focused on the collective attainment of that result. When a superior plays subordinates off against each other, overstimulates rivalry in other ways, or acts competitively with his subordinates, he forces them to divert energies from competition with other organizations into interpersonal rivalry. Less attention is focused on problems which the organization has to solve. In addition, the subordinates become defensive, or destroy cooperative possibilities by attacking each other, or maneuver for the favor of the boss. The most intensely intraorganizational rivalry is stimulated, the more acute the problem of company politics becomes.

Error #4: Failure to Anticipate the Inevitable

Many experiences in life are painful to people. Some, like aging and its accompanying physical infirmities and incapacities, are the lot of everyone. Others are specific to a man's work life, e.g., failure to obtain an expected promotion or the prospect of retirement. We can speak of such painful experiences as *psychological injuries*.

Such injuries are inevitable. Yet there is little in our 287 cases to indicate that companies recognize their inevitability and have established methods for anticipating or relieving them. The result is that those who are hurt in this manner have considerable hostility which is repressed or suppressed.

Error #5: Pressuring Men of Limited Ability

The characteristic and futile way of trying to deal with men of limited ability is by frontal assault. Repeatedly the senior executives attempt to persuade a rigid person to stop being rigid, exhort a dependent person to become independent, or cajole an impulsive person to gain better self-control. Although the executives may know in their minds that the subordinate is inflexible or unable to accept responsibility or assume initiative, they tend to act as if they could compel or stimulate him to change.

Error #6: Misplacement

Despite the plethora of psychological consultants, assessment and rating scales, and a wide-ranging literature on promotion, there is little indication in our cases that careful assessments are regularly made to indicate a man's limitations or predict his inability to carry greater responsibility.

Changes that will increase employee safety must take place at various levels of the organization, the most important of which is upper management. Levinson's discussion of upper-management problems is most relevant to the institution of human-error controls. Top management must do some things differently. There will be resistance, often caused by Levinson's six errors.

CHANGE

Figure 18-1 shows the influences needed to bring about change in an organization. As shown in the figure, there must be a stimulation of the power structure to institute change. That stimulus normally will be a safety professional who reads his or her role, correctly, as the role of a corporate change agent.

Figure 18-2 shows how individuals in the organization respond to change. As the model indicates, the more we can clarify what the change will mean for the individual, the more he or she has some control of the change, and the more he or she trusts the initiators of change, the more likely it is that he or she will be to accept change.

Figure 18-3 looks at change from the lower organizational level. At the employee level, change acceptance is a function in part of the employee's evaluation of that change: Does it make sense to him or her?

Management's reaction to change determines change's success. When upper management "buys in" to the changes, it ensures success. Upper management then expects lower-level managers and employees to "buy in" also. If upper management expects certain behaviors from lower levels, those behaviors will occur. Conversely, if upper management does not expect the behaviors, they will not occur. This simple fact has been documented over and over in safety and in nonsafety areas. It is known as the *Pygmalion effect,* or the self-fulfilling prophecy. What this means to safety is simply this: when changes are instituted, if top management expects those changes to be accepted and put into effect by lower-level managers, they will be. If top management (and middle management) provides only lip service to the change, then obviously the entire exercise will be a total waste of time.

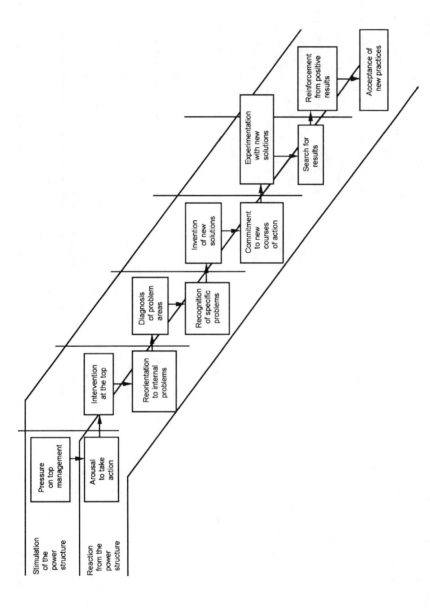

Figure 18-1. Influences that bring about change in organizations. (Adapted from G. Dalton, Influence and organization change. Cambridge: Harvard University, 1970. Photocopied.)

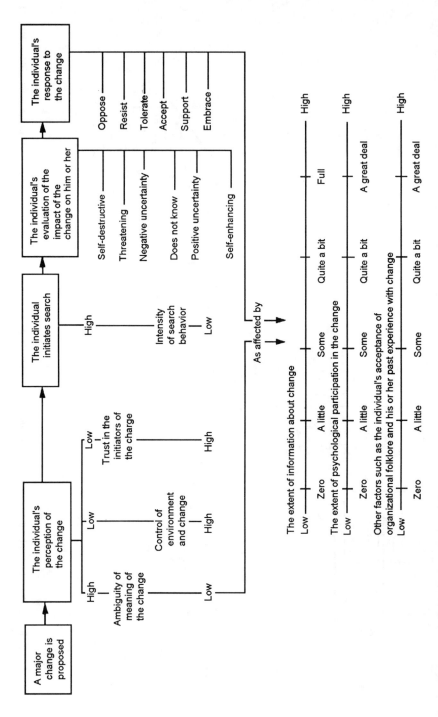

Figure 18-2. An individual's response to change. (Adapted from W. Bennis, in G. Lippitt, *Visualizing Change*. La Jolla, CA: University Associates, 1976.)

UPPER-MANAGEMENT CHANGES

We must ensure that upper management understands the thinking and concepts behind the safety controls (changes) to be instituted; namely, the human-error (HER) model and the major concepts discussed earlier. Margolis provides a series of guidelines for human-error reduction with which management ought to become familiar:[4]

1. Organizations should have as part of their philosophy a commitment to safety.
2. Intergroup competition for productivity, if safety is not included in the competition, will increase probability of accidents.
3. The leader of the work group by his style of leadership affects the accident potential among members of his work group. Though what makes for good leadership is complex leadership training for supervisors of high accident risk workers is an important tool in reducing accidents.
4. Irrelevant information while a man is operating a machine should be minimized to reduce risk of error or accident.
5. Whenever possible, self-paced work should be designed into work operations as opposed to machine-paced work. Self-pacing allows the intervals between task activities according to his own individual capabilities and needs, thus reducing stress, fatigue and probability of error or accidents.
6. In situations where vigilance and/or fast reaction time is an important part of the job task, the use of multiple cues, e.g., both visual and auditory signals, can enhance accurate and fast worker reactions and thereby reduce probability of error or accidents.
7. Workers should be trained in work practices which are as close as possible, if not identical to, the work practices that they will perform on the job.
8. Workers should be given immediate and specific feedback as to how they performed on the training task.
9. Proper responses to emergencies should be defined and practiced to the point of being overlearned.
10. To instill safe work practices in a given worker, a supervisor, or other member of an organization, must reward safe practices when they occur and be sure that unsafe practices are not rewarded. The key to accomplishing this is the correct identification of what is rewarding to that particular worker.

---

4. Reprinted by permission of the publisher, from The Company Personality: What It Is, How It Got That Way and What To Do About It, March 1959. © 1959 by American Management Association, Inc., All rights reserved.

11. To encourage workers to work in the safest possible manner, behaviors which approximate the desired working pattern should be rewarded first and, once these are learned, only those behaviors which come closer and closer to the desired end-state should be rewarded.
12. Punishment must be paired with the behavior at the time it occurs and must not be a rare event in itself. Threats of loss of pay, fines, loss of job and verbal abuse are more effective means of applying punishment than the possibility of accidents they seek to prevent.

Several areas have been suggested for upper-management's direct involvement: climate analysis, barrier-removal analysis, drug-abuse control, development of inverse performance standards, leveling analysis, improvement of selection and screening, and physical examination programs, some of which are discussed below.

Climate Analysis

Saul Gellerman wrote of analyzing a company's "personality" as follows:[5]

> There are a number of methods that can help to define a company's personality, but none of them can lead to a permanent definition—simply because company personalities keep changing. As economic pressures from without and the personal line-up within are altered, the company personality adjusts to them with shifts of attitude. Therefore, if analyses of this kind are to be made at all they should be repeated periodically—every three or four years—lest they become obsolete.
>
> Generally speaking, there are five steps to analyzing company personality: (1) identifying the company's pacesetters—the men whose attitudes count; (2) defining their goals, tactics and blind spots; (3) defining the economic challenges facing the company; (4) reviewing the company's history, particularly in terms of the careers of its leaders; (5) integrating the total picture—which in this case means to extract the common denominators, rather than to add up all the parts to get a sum.
>
> 1. *Identifying the Pacesetters.* This is not as easy as it sounds, because the real pacesetters are not necessarily in nominal command. A company pacesetter is defined by the reactions of other people to his ideas, rather than by his position in the company.
>
>   People seek out pacesetters to ask their opinions. These opinions are seldom questioned—except by other pacesetters. The working habits of a

---

5. Reprinted from S. Gellerman, *The Uses of Psychology in Management,* New York, Collier, 1970.

pacesetter are unconsciously copied by others in the company, sometimes to the point where they become standardized. In time, the pacesetters are surrounded by "disciples"—usually rising young men who hitch their personal fortunes in the company to the pacesetter's star. People whose views are incompatible with those of the pacesetters tend to lose influence or to leave the company. The pacesetters are exponents of the points of view that survive.

Even in a large company there are rarely more than a few real pacesetters. Therefore the analysis of a company personality begins with a search for a few men rather than a mass survey of every executive on the payroll.

2. *Goals, Tactics and Blind Spots.* This is the most important—and most difficult—step in the process. The temperament of the pacesetters should be assayed through both direct observation and a review of their records.

There are two key questions here. The first concerns the pacesetter's self-concept, especially the way he thinks about his role in the company. Does he see himself as a trail blazer, or a juggler with a dozen decisions in the air at any given moment, or a guardian of the stockholder's money—or what? The second important consideration is his underlying motive: what is he after personally? Does he want to ensure his own future security, to become prominent, to accumulate wealth, to wield power or to have people like him? Usually several motives are present, some more strongly than others. The predominant ones can have a powerful influence on the company personality.

The answers to these kinds of questions will usually reveal underlying patterns of consistency. In the last analysis, the way in which the pacesetters think about the company bears a definite, traceable relationship to the way in which they think about themselves.

3. *Economic Challenges.* As far as the company personality is concerned, the actual economic problems that it is objectively faced with are not nearly so important as the pacesetter's perceptions of these problems. An impartial economist and the company's pacesetters will not necessarily agree on a definition of the company's markets, competition, opportunities or resources. Whether the company's strategies turn out to be realistic or quixotic will depend, of course, on how accurately the pacesetters have estimated the situation.

It is here that a crucial element of the company personality appears—its adaptability or capacity to make realistic choices regardless of its inclination and wishes. A company with a rigid personality will adhere to its habitual ways of operating come hell or high water (and sometimes the water gets too high). A company with an adaptable personality will discard a tradition when it becomes untenable or when a more profitable policy is at hand.

Some of the best examples of this can be seen in connection with diversification and new-product development. The old established firms which cherished their reputation for reliability and banked everything on the perpetuation of the public's taste for craftsmanship have, unfortunately, been driven to the brink of bankruptcy fairly often in the last decade. Their pacesetter's self-image of being a guardian of quality was painfully won and even more painfully relinquished. As far as fashions in pacesetters go, this seems to be the day of the light-footed, far-sighted genius who can smell a shift in consumer tastes before it happens, and jumps from the old bandwagon to a new one in plenty of time to make the most of both.

4. *Company History.* To trace the history of a company personality one needs to be more concerned with ideas and people than with production or sales figures.

To begin with we need to know about the sources of new (or potential) pacesetters and of new ideas. Nepotism, which is not necessarily an evil but is usually viewed with a jaundiced eye by outsiders, often enters the picture. Since some families are notoriously lacking in harmony while others manage to have its [sic] members see eye to eye about nearly everything, no sweeping generalization can be made as to whether nepotism breeds ideas or leaves them to stagnate.

However, the men who are singled out for advancement by the powers that be will usually be in sympathy with prevailing ideas about the company's business. The nonconformist is often weeded out or squelched. Perhaps this is as it should be, and perhaps it isn't, but in any case it creates a danger that the key ideas by which a company is managed will become inbred, uncriticized and obsolete.

Therefore any source of pacesetters in the company personality has to be considered as part of the company's resources. There is often a tendency to seek new managerial talent among men who share a certain common background. For example, one large Eastern firm regularly chose about 90 per cent of its management trainees from recently discharged servicemen who had no previous job experience and were graduates of a particular denominational college. Until this was pointed out to them, the company's executives were unaware that nearly all of their prospective successors were running to a single type!

"Inbreeding," either through families or through similar backgrounds, is not necessarily unwise. It is often unavoidable, and in such cases it is valuable for management to develop some insight into the special advantages and limitations imposed by their "homogenized" background.

5. *Finding Common Denominators.* Out of all this data there will usually emerge a fairly consistent set of patterns that reveal the basic attitudes of the pacesetters toward the company's business. The areas of agreement among the pacesetters will set the norms of behavior and thinking to which other members of the company are expected to conform. In some compa-

nies the emphasis will be on punctuality, "correctness" and restraint; in others the keynotes will be informality and "togetherness," and so on.

Areas of emphasis and deemphasis will become apparent. In some companies the production department is at the center of everything and other departments seem to exist largely to service and assist the favored few. In others the sales department dominates the scene and can just about dictate the kinds of actions the other departments can take. Even where balance exists there will often be underlying strains pulling in one direction or another.

Some companies will be found to be at cross purposes with themselves. (One is tempted to talk about a "split company personality.") Some symptoms of this are: vacillation between opposed courses of action, too much "study" and analysis of proposed moves before approving or rejecting them, and repeated makeshift "solutions" which fail to strike at the heart of long-standing problems.

WHY COMPANY PERSONALITIES SHOULD BE ANALYZED—AND WHY THEY SHOULDN'T

Merely because a company presumably has a personality of its own is certainly no reason to attempt to analyze it. The real essence of a company's human resources is as elusive and easy to misjudge as the real motives of the individual human being. Even when it is caught there is no guarantee that anything constructive will come of it.

But insight into the company personality can, in certain situations, be of far-reaching value and not just an academic exercise. The two prerequisites for a useful analysis of this type are dissatisfaction with the company's demonstrated ability to meet its problems and a plausible basis for suspecting that unexamined group attitudes, rather than individuals or policies, are the culprits.

When things go wrong there is an all-too-human tendency in companies and elsewhere, to "solve" the problems by causing a few heads to roll. Many a sound, valuable executive has been sacrificed this way to pay for mishaps that could not have been avoided under the conditions imposed on them by company-wide blind spots and miscalculations. This is a tragic waste. Finding scapegoats may ease a few consciences but it isn't nearly as effective, in the long run, as finding causes.

The analysis of a company personality should never be undertaken with a commitment to alter or "improve" what is found. There are often practical reasons why changes, however desirable they might be in theory, are out of the question. More important, some companies cannot change because of the character of their pacesetters: to induce such a company to alter its basic thinking is merely to court unnecessary stress and eventual disaster.

The analysis should not assume, in other words, that the company personality is "sick," because a "cure" may be impossible and because such an assumption may only obscure hidden strengths. These strengths are as much a target of investigation as weaknesses.

There are five kinds of benefits to be derived from surveys of this kind. The most obvious concerns executive selection and placement. Incompatibility between the individual executive and the company personality is at the bottom of more short-lived placements than many people realize.

Second, an analysis of the company personality should precede the planning for a management development program. This will help to pinpoint the kinds of thinking and activities that need to be fostered. More important, it will help to throw light on the attitudes—and people—that might hamstring the program by opposing or undercutting its qualities.

Third, the analysis can help spotlight obsolete or inefficient policies which are clung to because everyone assumes that they are immutable or because people have lost sight of the fact that they are policies and regard them as inescapable conditions under which the business must operate.

Fourth, learning about the company personality can help to delineate the limit of the company's tolerance for new ideas or new faces. In a market-conscious era that seems to demand an unending process of revamping, many companies commit themselves to costly programs for which they have no real stomach when the full ramifications of the program become apparent.

Fifth, the analysis can help redirect the company toward areas where its own special balance of strengths and weaknesses can be applied to maximum advantage. Some companies try to "be all things to all men" and artificially equalize their emphases despite a natural inequality of talent and interest. For example, the naturally conservative company is probably making a mistake when it starts gambling on costly promotional schemes that may or may not pay off—unless, of course, it has succeeded in changing its basic personality beforehand.

This brings us to the ultimate question of what can be done about a company personality that seems to be out of step with its time. Although it is always possible to change the company personality, it is not, as we indicated earlier, always feasible to attempt it. Change is brought about by training, by improving internal communications and/or by bringing in new personnel. We should not forget that change is also brought about by time and that it is sometimes less damaging to wait for a natural evolution than to turn the company topsy-turvy all at once.

In Chapter 5 safety-program climate was discussed briefly. Obviously, both corporate and safety-program climates need to be looked at before changes are decided upon. This analysis ought to be undertaken by upper management, although it is entirely possible that the safety staff will be involved. Chapter 5 also discussed several theorists who have written on corporate climate. For the reader interested in learning more about climate, Rensis Likert's book *The Human Organization* is a valuable resource. It contains a climate questionnaire, which might be of value to any safety professional looking closely at the climate of his or her organization.

A quick, superficial climate reading of an organization can be obtained by filling out the form in Figure 18-3 after reading Gellerman's discussion of corporate personality and reviewing the climate factors explained in Chapter 5. If the form is filled out by people at various organizational levels, the evaluation will increase in validity.

Barrier-Removal Analysis

Closely allied with climate analysis is the concept of barrier-removal analysis. McGregor suggested that the primary role of the Theory Y manager was to remove the barriers that exist between a worker and his or her work. Such barriers might be lack of supplies, equipment problems, management goofs, red tape, or problems coordinating with other departments.

In barrier-removal analysis, one upper-management activity might be to see how lower-level managers are carrying out barrier removal. This might be done through observation or by asking lower-level managers to keep an activity log for a period of time and then categorizing those activities, for example, supervising, fire fighting, barrier removing, and so on.

Inverse Performance Standards

The traditional way to make performance appraisals is to have the boss evaluate the subordinate. When inverse performance standards are used, the subordinate evaluates the boss. Figure 18-4 shows one possible way of doing this. Subordinates fill out the form and submit it anonymously to a third person, who keeps the responses confidential and provides the boss with a summary of all responses.

Leveling Analysis

The rating form in Figure 18-4 asks some questions that are pertinent to the leveling analysis. Since the interpersonal relationship between boss and employee is an important factor in the employee's performance, it seems worthwhile to attempt to analyze this relationship. Given the proper training in interpersonal relationships, a supervisor might be expected to maintain a rela-

Organization _____ Date _____

Assessment by _____

Safety-Program Climate

Lively |—|—|—|—|—|—|—|—| Negligent

Corporate climate factors   Good   Poor

Factor		
Confidence and trust		
Subordinate interest		
Understanding of problems		
Training and helping		
Teach how to solve problems		
Giving of support		
Information dispersal		
Opinions sought		
Approachability		
Recognition		
Summary		

Corporate climate requirements

Factor		
Goals		
Communications down		
Department goals		
Interdependency		
Participation		
Freedom to work		
Expansion		
Delegation		
Innovation		
Fluid communication		
Stability		

Corporate personality

The pacesetters Who _____
               Type _____
               Self-concept _____
               Motives _____
Economic challenges _____
Our history _____
The common denominators are _____
_____
_____

Figure 18-3. Climate-assessment form.

Department

Note: Do not sign your name. Your boss will not see this sheet. He or she will receive a summary of all responses from this department.

Consider your boss and how he or she performs compared to your expectations of him or her.

Does your boss:	Better than I would expect	Worse than I would expect
	10	1

Know you?_____
Understand you?_____
Know what your needs are?_____

Write any comments here you wish

Back you?_____
Listen to you?_____
Talk to you?_____

Write any comments here you wish

Allow your input?_____
Ask for your ideas?_____
Use your ideas?_____

Write any comments here you wish

Remove any barriers in your way?_____
Have enough influence with his or her boss?_____
Have enough influence with other departments?_____

Write any comments here you wish

Talk down to you?_____
Treat you as a child?_____
Treat you as a subordinate?_____

Write any comments here you wish

Figure 18-4. Inverse-performance-standards rating form.

tionship that is motivational. Training in transactional analysis (TA) is particularly helpful for this. Once trained in TA, supervisors might be expected to maintain an adult-to-adult relationship with each employee.

The last questions on the form in Figure 18-4 will help define the relationship. Periodic contacts with lower-level employees by representatives of upper management are informative also. The important thing is that upper management care enough to find out what kinds of relationships exist.

Other Upper-Management Tasks

One other upper-management task is of utmost importance. That task is to ensure that all lower-management tasks and safety-staff tasks are carried out. A system of accountability is a must to ensure that this is done.

In the next section we discuss the human-error-reduction tasks of lower management. Whether or not those tasks are carried out regularly and routinely totally depends on upper management's interest, and the systems it has devised and installed for finding out whether or not lower management is carrying out its defined tasks.

LOWER MANAGEMENT

Lower-management involvement in human-error control is crucial, as it has always been in occupational safety. The greatest number of things that might be done to control human error must be done by lower management. Eliciting and maintaining satisfactory, safe performance from employees is lower management's job. It is a continuing job, and a difficult one. To give the lower manager the necessary tools, behavioral science has investigated many motivational techniques. Out of these investigations has come the consensus that the way to elicit the proper response from employees is to fulfill their psychological needs.

To use this motivational approach, we have to assume at the outset that the supervisor knows what the employee needs at any particular time. This is often an erroneous assumption. Figure 18-5 shows a simple form that has been used in training sessions for both supervisors and employees for many years. Supervisors and employees are asked to rate what they believe employees want most from their jobs. The filled-in columns on the right show typical ratings made by each group. Supervisors usually rate high wages as number 1, job security as number 2, and promotion in the company as number 3. Employees usually rate appreciation of work done as number 1, feeling of belonging as number 2, and help on personal

Line Management Roles 291

	Rating by Employees	What Supervisors Think Employees Want	
1. Help on personal problems	#3	#9	
2. Interesting work	#6	#5	
3. High wages	#5	#1	
4. Job security	#4	#2	
5. Personal loyalty of supervisor	#8	#6	
6. Tactful disciplining	#10	#7	
7. Full appreciation of work done	#1	#8	
8. Feeling of belonging	#2	#10	
9. Good working conditions	#9	#4	
10. Promotion in the company	#7	#3	

Figure 18-5. What do employees want most from their jobs? Rating form for employees and supervisors.

problems as number 3. There could not be wider differences between the perceptions of the two groups.

Since it is generally agreed that supervisors can motivate workers by satisfying their current needs, it is imperative that supervisors be able to identify those current needs and provide a job situation in which the needs can be satisfied. As Figure 18-5 shows, supervisory perceptions of employees' needs tend to be rather poor. Other studies tend to bear this out. Figure 18-6 shows similar results from Handy's research, and Table 18-1 shows some results from Kahn's research.

The first task, then, seems to be to improve supervisors' perceptions of employees' needs. The second is to effect changes that satisfy employee needs. First-line supervisors might well need specific instruction and training in two distinct areas if they are to be effective in human-error reduction: (1) they must be taught how to assess and then meet employee needs, and (2) they must be taught specifically what tasks they must do in order to reduce human error. For the first area, training in transactional analysis seems particularly appropriate, for it teaches how to assess needs and achieves leveling at the same time. For the second area, the specific tasks and activities covered in this chapter might be helpful.

Worker Safety-analysis form.

1. Of the employees,

| 35% | | say their supervisor understands employees' problems well.

2. Of the supervisors,

| 95% | | say they understand the employees' problems well.

Nevertheless, of these supervisors,

| 51% | | say their managers understand the supervisor's problems well.

3. Of the managers,

| 90% | | say they understand the supervisor's problems well.

But among these managers,

| 60% | | say their superior understands managers' problems well.

Extent to which superiors and subordinates agree as to whether superiors tell subordinates in advance about changes.

	Top staff says as to own behaviour (percentage)	Supervisors say about top staff's behaviour (percentage)	Supervisors say as to own behaviour (percentage)	Employees say about supervisor's behaviour (percentage)
Always tell subordinates in advance about changes that will affect them or their work	70 } 100	27 } 63	40 } 92	22 } 47
Nearly always tell subordinates		36	52	25
More often than not tell	30 }	18	2	13
Occasionally tell	—	15	5	28
Seldom tell	—	4	1	12

Figure 18-6. The communication gap between employees and supervisors. (Adapted from C. Handy, *Understanding Organizations.* Middlesex, Eng.: Penguin Books, 1976.)

Table 18-1. What subordinates want in a job compared with their superiors' estimates (percentages)

Variables Measured	As Subordinates Rate the Variables for Themselves	As Supervisors		As General Supervisors	
		Estimated Subordinates Would Rate the Variables	Rated the Variables for Themselves	Estimated Supervisors Would Rate the Variables	Rated the Variable for Themselves
Economic variables					
Steady work and steady wages	61	79	62	86	52
High wages	28	61	17	58	11
Pensions and other old-age security benefits	13	17	12	29	15
Not having to work too hard	13	30	4	25	2
Human-satisfaction variables					
Getting along well with the people I work with	36	17	39	22	43
Getting along well with my supervisor	28	14	28	15	24
Good chance to turn out good-quality work	16	11	18	13	27
Good chance to do interesting work	22	12	38	14	43
Other variables					
Good chance for promotion	25	23	42	24	47
Good physical working conditions	21	19	18	4	11
Total Number of cases*	2499	196	196	45	45

Source: Adapted from D. Kahn, Human relations on the shop floor, in *Human Relations and Modern Management,* edited by Hugh-Jones, 1958.
*Percentages total over 100 because they include three rankings for each person.

What Determines Performance of Supervisors

What makes supervisors act on their assessments of what must be done to reduce human error? What determines whether or not a supervisor will, in fact, carry out the tasks once they are defined?

Since the beginning of the industrial safety movement, it has been the fundamental belief of safety professionals and managers at all levels that if anything is to be accomplished, it will be accomplished by the supervisor. Unfortunately, adherence to this belief has not gotten the job done. Although we have made substantial progress in safety, the real progress was made during the early years. In recent years, progress has ceased.

One reason for this lack of progress is an inability to transform our fundamental belief about supervisory responsibility into supervisory action. Most supervisors in American industry do not do a good job of carrying out safety programs. There are many reasons for this failure, including the following:

- Management has seldom communicated clearly what should be done in safety. Management says, in effect, "It's your responsibility, now go do something about it."
- Management almost never follows up to see if the supervisor has carried out a safety plan. If anything, management may look at a safety record.
- Management almost always communicates clearly what should be done in other areas, such as quantity and quality of production. Moreover, management almost always measures supervisors' performance in those other areas.

Why, then, should any supervisor spend time and effort on safety? Management can change all this by:

- Deciding what performance it wants from its supervisors. This can include such things as "care" contacts, "care" walks (such as inspections), and "care" searches (such as investigations using some new techniques).
- Adopting a system that shows whether or not each supervisor is carrying out this performance.
- Setting criteria for acceptable performance.
- Rewarding supervisors according to their performance.

This philosophy is well illustrated by Edward Lawler and Lyman Porter in their model of supervisory performance (Figure 18-7).

LOWER-MANAGEMENT TASKS

As we have stated, management elicits supervisory performance by (1) deciding what supervisors should do, (2) training them (if necessary) to do it, (3) measuring their performance regularly to ensure that they are doing it, and (4) rewarding them accordingly. The first step is to decide what tasks are to be accomplished. A number of tasks have been identified for lower management related to human-error control. Included are analysis of the arousal state, attitude analysis, the use of behavior modification, use of information on biorhythmic

Line Management Roles

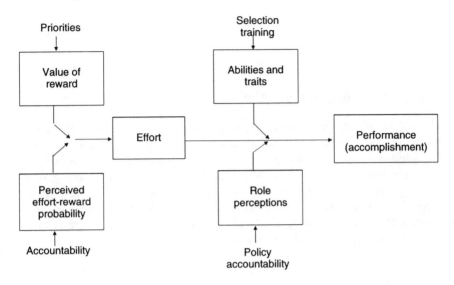

Figure 18-7. Supervisory model. (From E. Lawler and L. Porter, *Managerial Attitudes and Performance*. Homewood, IL: Richard D. Irwin, Inc., 1968.)

state, crisis intervention, the fatigue and boredom check, force-field analysis, group analysis, information-processing overload check, job redesign if indicated by the life-change-unit (LCU) check, task analyses, and training. Many of these tasks and a few others might be included in a concept called work-safety analysis.

Figure 18-8 is a worker-safety-analysis form. Some items can be left off the form if the information is not available, although LCU information might be available to a supervisor who knows his or her people. Personality type and accident risk are optional, and might not be filled in by many. These items may be less valuable, from a safety standpoint, than the other items on the form. The value analysis is quite arbitrarily decided by the supervisor, and might or might not be useful.

Current motivational analysis is the most valuable item on this form. The supervisor should look at each item listed and determine what motivational pull it will have on the employee. Included are most of the important determinants of employee performance we have discussed in this book.

The supervisor should consider current job assignment in terms of the load it places on the person. The last item on the form is force-field analysis; the supervisor may choose to perform a small force-field analysis to determine the current pulls on the employee. The entire worker-safety analysis might lead to some disposition, as shown at the bottom of the form.

Obviously, the supervisor who fills out this kind of form will need considerable training in order to understand the concepts involved. The purpose of worker-safety analysis is to help the worker; its intent is to spot the causes of human error before an accident can occur.

Dealing with Human-Error Causes

The bottom of the form (Figure 18-8) offers several dispositions. If no real problem is unearthed and no action is indicated, the "OK" box is checked. Other disposition options are to discuss the analysis with the person, as in the case of a relatively minor problem; to send him or her to receive additional training, or to give it yourself; to administer crisis intervention, in the event that the analysis reveals something crucial and critical that must be dealt with; to draw up a contract for behavioral change; or to use behavior modification. Some of these dispositions require some additional comment.

Training

This is one of the simplest solutions, and usually one of the most ineffective. It assumes that we have identified a lack of knowledge or skill. If this assumption is correct, training is the proper solution and will be effective. If, however, the assumption is not correct, that is, the problem is not a lack of knowledge or skill, a different solution is indicated.

Crisis Intervention

Obviously, crisis intervention is indicated only if a severe, immediate problem exists.

Behavior Modification

Behavior modification is not new, and it is relatively simple to understand. The basic process involves systematically reinforcing positive behavior while ignoring or penalizing unwanted behavior. The end result is the creation of a more acceptable response to a given situation. The technique concentrates on a person's observable behavior and not on its underlying causes. There are two primary approaches used in behavior-modification programs. One is the attempt to eliminate behavior that detracts from organizational goal attainment, and the other is the learning of new responses. The safety professional would like to eliminate unsafe acts. This elimination process is called *extinction.* Extinction may be accomplished through either punishment or withholding of the positive

Worker-Safety Analysis

Name _____ Date _____

Long-term analysis

  Biorhythmic information: dates to watch: _____
  LCU information: approximate units accumulated now: _____

Personality and value analysis

  Personality type _____
  Accident risk _____

	Key importance	No importance
Value of work		
Value of safety		

Current motivational analysis   Turn ons   Turn offs

  Peer group _____
  Me (boss relations) _____
  Company policy _____
  Self (personality) _____
  Climate _____
  Job-motivation factors _____
    Achievement _____
    Responsibility _____
    Advancement _____
    Growth _____
    Promotion _____
    Job _____
  Participation _____
  Involvement _____

Current job assignment   High   Low

  Pressure involved _____
  Worry or stress _____
  Information processing need _____
  Hazards faced _____
  Other _____

Force-field analysis

Pulls to safety
  ↑
  ↓ _____
Pulls away from safety

Current assessment:

❏ OK   ❏ Discuss with worker   ❏ Training   ❏ Crisis intervention
                    ❏ Contract
❏ Crisis intervention   ❏ Behavior modification

Figure 18-8. Worker-safety-analysis form.

reinforcement. There are two difficulties associated with punishment, or negative reinforcement: (1) backlash, or an undesirable reaction; and (2) extinction that is of short duration. Punishment may cause the person to suppress the behavior only temporarily; it may emerge again when the negative consequence no longer follows. Punishment teaches the worker not to perform the unsafe act in our presence. He or she will wait until we are gone.

Extinction of behavior may also be accomplished if we withhold positive reinforcement. If customary rewards for the action in question are no longer granted, then extinction will occur. For example, if we fail to laugh at the unhumorous stories told by the office joker, his or her reward is not forthcoming—we have withheld positive reinforcement. If reward is repeatedly withheld, the bad jokes eventually will be eliminated. Since the use of punishment may only temporarily suppress and not totally extinguish behavior, the logical alternative is to withhold positive sanctions.

The second major goal of behavior modification is to create acceptable new responses to an environmental stimulus. As in eliminating behavior, reinforcement techniques play the vital role in developing new behavioral responses. Directly rewarding desirable responses is the best way of stabilizing intended behavior patterns. Eliminating undesirable behavior without providing a new substitute pattern leaves the worker open to learning another undesirable set of responses. The substitution of new, desirable responses for unwanted behavior is the overall purpose of behavior modification. The goal is to leave the individual with new or modified behavior patterns in place of behavior that was deemed unwanted or unneeded.

Learning theory carefully explains that reinforcement must be applied according to a detailed, systematic plan. The reinforcement schedule that is used depends on the nature and degree of behavior that the person presently demonstrates.

If the worker is presently not performing as we wish (not exhibiting the desired behavior), obviously no reinforcement is possible. In this case we have two possible choices. First, we can inform the individual what response would lead to a reward. ("If you perform in the following manner, you will be given a raise.") The employee is now aware of what response must be made to obtain the reinforcement, and the desire to receive the reinforcement will cause him or her to act. The second approach to reinforcement is the shaping of new behavior from existing behavior. In this case, the desired pattern does not exist (and, therefore, cannot be reinforced). To elicit the new behavior, a reward is given for responses that closely approximate it. Modification, or shaping, is accomplished by discriminately reinforcing behavior that is close to the desired behavior. A closely controlled reinforcement schedule is applied when the wanted end behavior appears.

Reinforcement schedule merely means how and when we choose to reinforce behavior. Many types of reinforcement schedules are available. Most types can be classified as (1) fixed or variable and (2) interval or ratio. A fixed schedule (reinforcement each time the desired behavior occurs) might be employed until the pattern is at an acceptable level. Later, rewards may be administered on a variable (not every time) basis. The decision on whether to used a fixed or variable schedule and whether this schedule should be based on time (interval) or on the number of performances (ratio) must be made carefully. For example, production jobs that can be performed and measured by each piece produced are easily adapted to ratio schedules. On the other hand, administrative positions are better suited to interval scheduling. Generally, when possible, variable schedules of reinforcement should be used because they tend to provide the greatest resistance to extinction.

Since we want each worker to perform safe acts indefinitely on into the future (with great resistance to extinction), a *variable-ratio* schedule of reinforcement is best. This merely means that we do reward the worker for the safe behavior, but not each time he or she exhibits it. We might reward the worker after 6 safe acts, or after 20—the number varies. The reward is not given after varying amounts of time have passed, but after varying numbers of safe acts have been performed. A variable-ratio schedule of positive reinforcement is the exact schedule that makes gambling so fascinating and that builds the compulsive gambler. The gambler knows there is a reward coming, but he or she does not know when. Any person who has tried the one-armed bandits knows how a variable-ratio schedule of positive reinforcement works.

Systems of rewards currently being used in industry and other types of organizations can be adapted to a behavior-modification program. Money, status, promotion, public recognition, and personal praise are all valid reinforcers of behavior. In addition, much research has been done using tokens as reinforcers. While to date most of this research has been with retarded children, psychotics, slow learners, and juvenile delinquents, there are indications that it might be used more and more with "normal" people. Tokens could perhaps be used to reinforce safe behavior. Trading stamps are already used in some safety contests.

To administer a reward system properly, we must first determine exactly what behavior to reinforce. First, we must decide exactly what we want the person to do. Second, we must determine how we will reward the person for achievement. The assumption that the paycheck reinforces job performance is not necessarily valid. Superficially, we might think of salary as a fixed-interval schedule of reinforcement in the sense that money is a reinforcement that is delivered every 14 or 30 days. Actually, however, money reinforces only the behavior of accepting the paycheck. This performance is reinforced every time it occurs. If salary attainment were more related to job behavior, it would be a better reinforcer. This relationship has been achieved in many companies, particularly in the form of management compensation, where the bonus (compensation) has been related to safety achievement. Generally this approach has worked. Its weakness is that the reinforcement comes only once a year. Learning theory holds that the reinforcement should immediately follow the performance. Salary increments granted according to attainment of specific predetermined objectives would incorporate a ratio schedule that would provide more direct reinforcement than salary administered on an interval basis.

Here is an example from a production situation. Suppose that a worker's job consists of installing a taillight on an automobile as it passes along the assembly line. We might attach a small device similar to a taxi meter to the machine the worker employs. This would ring a bell after the worker had made a sufficient number of responses, and the amount earned would flash on a screen before him or her. Under a setup like this, the worker might have to be pulled away from the line for coffee breaks.

Contracts

A contract can also effect behavior change in an employee. A contract is an agreement to do something about something; an agreement between the supervisor and the employee. Contracts can be established to change behavior, to change feelings, or to change physical conditions such as high blood pressure and obesity. In a work situation, contract are primarily used for the purpose of changing behavior. According to the book *The OK Boss* by Muriel James, there is a five-step process for making contracts:

1. The establishment of a goal. James suggests that the employee ask: What do I want that would enhance my job?
2. The definition of what needs to be changed to achieve the goal. What would I need to change so that I can reach my goal?
3. The determination of how much the person is willing to do to achieve that goal. What would I be willing to do to make the change happen?
4. The determination of measurement and feedback needed to accomplish the change. How would others know when I have affected the change?
5. Other considerations what pitfalls are there in the way? How might I sabotage or undermine myself so that I would not achieve my goals?

Muriel James states that each of the five points should be discussed when a contract is made. Written answers to each of the questions are also helpful.

The contract is an excellent management tool for dealing with employees. The employee is allowed to participate in what is going on, which is always preferable to authoritarian enforcement.

Group Approaches

Up to this point, we have discussed approaches to human error-reduction that involve only the worker and the supervisor. Group approaches might also be used. The power of group pressure has been shown many times in controlled experiments.

Line Management Roles

Figure 18-9 illustrates results of a test of high school graduates, all men, with good eyesight. They were to tell whether two lines were the same length or not. Most of the pair of lines were noticeably different in length. When tested individually, the men were almost 100% correct in judging which line was longer. But then they became the unsuspecting victims of group pressure. Each man in the experiment was asked to judge the lines again, this time in a group setting. All the other people in the group were conspirators who had been told to call out wrong answers. Each conspirator called out the wrong answer before the time came for the innocent victim to report how the line lengths looked to him. The first two columns on the left show that the victims were only slightly misled when one or two others who preceded them called out wrong answers. But when three or more others gave wrong answers, one-third of the victims gave in. They took the group's word, rather than report what their eyes told them. This demonstrates the *majority effect*. It is significant that a group of three others was as powerful as a group of sixteen for misleading the victims.

Some of the victims said later, after being told of the practical joke played on them, that the line lengths honestly seemed to change as they heard others in the group call out the wrong answer. A few of the victims deliberately gave answers they felt were wrong; they did not want the group to think them different. But most of the victims felt the group must be right because it was unanimous. Somewhat similar tests have been made for judging the size of a rectangle. In these tests, people tended to change their estimates to conform to estimates they were told were made by a group of 20 to 30 people.

Our safety program must help us to build strong work groups with goals that coincide with our safety goals. We can determine the strength of our groups by observing some characteristics of strong and weak groups, as discussed in Chapter 13. We can exert some influence, of course, but basically competence, maturity, and strength are established qualities. Therefore, the strength of the group depends mostly on whom we place in the groups initially. How can we make good choices?

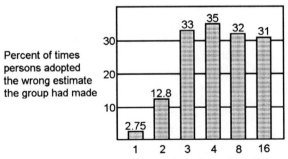

The numbers along the bottom of the graph are the numbers of conspirators that called out wrong answers.

Figure 18-9. Measurement of group influence on the judgments of individuals. (From D. Petersen, *Safety Management.* Aloray, Inc., Goshen, NY, 1988.)

First, how do we select and place employees initially? Are the three characteristics (individual competence, maturity, and strength) most important with respect to selection or with respect to placement in a work group? People who lack the three characteristics of strong-group members should be placed in work situations where they do not destroy a strong group, or where a strong group is not essential to the defined goals. Second, we might consider shuffling individuals into other groups. Sometimes the transfer of only a few individuals can make a major difference in several groups.

The fourth essential to a strong group is common objectives. In this area, management can do quite a bit. In safety it means doing a much better job of goal setting, motivating, and communicating so that the safety goal becomes the chosen group goal.

Other Approaches to Human-Error Reduction

In addition to the individual and group approaches suggested thus far, the supervisor also has the job-centered approaches of inspection, job safety analysis, task analysis, and job redesign. Most of these are somewhat traditional, or have been discussed elsewhere in this book.

THE WORKER ROLE

It was easier to define the worker's role in the safety system years ago; their role was to follow the rules, to work safely.

It is not simple today, and thank heavens. Today the worker's role is to be involved, to participate, in everything—input, decision making, running the entire safety program—as much as possible. We cannot define this role because it is to do it *all* if he or she chooses it. As much as the worker wants to do is encouraged; if he or she runs the safety system, so much the better! But with one caution: employee ownership does not let management off the hook. When the workers own the system, management at any level *cannot* abdicate their responsibility.

Hourly employees, if they will, can and should do it all—handle every aspect of the safety system—with total and complete management support and backing.

REFERENCES

Bennis, W. In G. Lippitt, *Visualizing Change*. La Jolla, CA: University Associates, 1976.

Carlson, N. The trouble with teams, in *Training,* August 1992.

Dalton, G. Influence and organization change. Photocopied. Cambridge: Harvard University.
Gellerman, S. Total safety culture, in *Professional Safety,* September 1994.
Gellerman, S. *The Uses of Psychology in Management.* New York: Collier, 1970.
Handy, C. *Understanding Organizations.* Middlesex, Eng.: Penguin Books, 1976.
James, M. *The OK Boss.* New York: Bantam Books, 1975.
Kahn, D. In Hugh-Jones, ed., *Human Relations and Modern Management,* 1958.
Lawler, E. and L. Porter. *Managerial Attitudes and Performance.* Homewood, IL: Irwin, 1968.
Levinson, H. Who is to blame for maladaptive managers? *Harvard Business Review,* November–December 1965.
Likert, R. *New Patterns in Management.* New York: McGraw-Hill, 1958.
Likert, R. *The Human Organization.* New York: McGraw-Hill, 1967.
Lindsay, G. and E. Aronason. *The Handbook of Social Psychology.* Reading, MA: Addison-Wesley, 1968.
Margolis, B. Psychological-behavioral factors in accident control. Paper read at the ASSE Professional Conference, Park Ridge, IL, 1973.
McGregor, D. *The Human Side of Enterprise.* New York: McGraw-Hill, 1960.
Moore, M. Behavioral technology and behavior change. *ASTME Vectors,* Vol. 5, 1969.
Petersen, D. *Safety Management: A Human Approach.* Englewood Cliffs, NJ: Aloray, 1975.
Petersen, D. The role of safety in the 1990s, in *Professional Safety,* June 1995.

Chapter 19

Role of Staff

Our final area of consideration is the role of the staff safety specialist, or safety professional, in human-error reduction. First, the safety specialist is a systems evaluator. This statement does not describe the actual duties of the staff safety specialist or tell us whether the specialist's job is full- or part-time. There is no one job description for the staff safety specialist. The duties that we perform are the duties we have set for ourselves. The job is self-defined, and the duties will vary, depending on the size of the organization, the number of locations involved, the operations themselves, the people above and below the specialist in the organizational hierarchy, the problems presently facing the company, the other safety staff people and specialists available, and the safety specialist's position in the organization.

In 1963, the American Society of Safety Engineers (ASSE) developed a logical approach to the development of the safety profession. The first phase was the identification of what a safety professional does. In that part of the study, the study group developed a paper entitled "The Scope and Functions of the Safety Professional," which was published in 1966 and remains a classic description of what a safety professional does within the organization. According to that document:[1]

> The major areas relating to the protection of people, property and the environment are:
>
> A. Anticipate, identify and evaluate hazardous conditions and practices.
> B. Develop hazard control designs, methods, procedures and programs.

---

1. Reprinted with permission from American Society of Safety Engineers, The scope and functions of the safety professional, Park Ridge, IL, 1966.

305

C. Implement, administer and advise others on hazard controls and hazard control programs.
D. Measure, audit and evaluate the effectiveness of hazard controls and hazard control programs. [See Figure 19-1.]

This scope and functions document reflects the most recent publication of the ASSE, in which there are only slight changes from the original document, indicating the direction of the profession. There are no changes in the four primary functions although there are changes in how some of them are defined.

For instance, in all earlier documents, our first role was to identify and appraise the management system to determine what system failures are present that could result in loss—in the physical, managerial, and behavioral environment. This role has now been expanded to anticipate, identify, and evaluate these changes.

In the earlier documents, the safety professional's role was to communicate to the line management what must be done to control loss, recognizing that the only persons who can control losses are in the line. In the recent document, our role is to implement, administer, and advise, removing the safety professional from the purely advisory, change-agent role to one of an active participant, allowing line managers the opportunity to abdicate their previously well-defined role.

In the earlier documents, the safety professional's role was to measure and evaluate progress, allowing a choice of which measures will do the most valid job of measuring safety system effectiveness. In the recent document, audit is defined

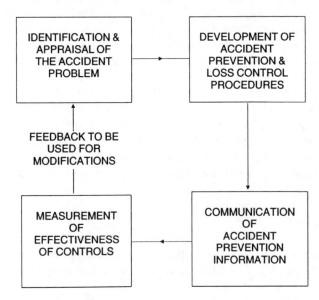

Figure 19-1. The functions of the safety professional. (Reprinted with permission from ASSE, The scope and functions of the safety professional, Park Ridge, IL: ASSE, 1966.)

as the measure (or one of the measures) to be used. This is no problem if the audit used correlates with the accident record over time (is, in fact, auditing what will get results); but if the audit is checking regulatory compliance primarily, it could be counterproductive, as is discussed in this book.

The safety professional's function, then, is to appraise what exists, develop improvement plans, communicate those plans to the line organization, and measure results. None of this changes in a human-error scheme of safety management. Number 1 becomes the identification and appraisal of causes of human error. Number 2 becomes the development of methods, procedures, and programs for the reduction of human error. Functions 3 and 4 remain the same. The safety specialist must carry out the same four roles for human-error reduction as for traditional safety programming.

First of all, the safety specialist must structure systems of measurement so that accountability can be fixed, and so that rewards can be applied properly to the right people and at the right time to reward or reinforce the desired behavior. Some of these systems might include the specific techniques discussed in the last two chapters. Second, the specialist must be a programmer—a person who oversees those aspects of the safety program that are not completely under his or her purview. This includes ensuring that safety is included in orientation, that safety training is provided where needed, that safety is a part of supervisory development, that things are done that help to keep the organization's attention on safety, and that safety is included in employee selection and in the medical program. To carry out these responsibilities, the safety specialist must install changes in the procedures and tasks of upper and lower management.

Third, the specialist must function as a technical resource. He or she must know how to investigate in depth; know where to get technical data; know the standards; and know how to analyze new products, equipment, and problems. Fourth, the safety specialist must function as a systems analyst, searching for "whys" and "whethers." Who might best fill this kind of job? This question has no easy answer.

In 1969, Bracey made a study of the effectiveness of safety directors. He attempted to test these hypotheses:

1. Safety directors who are more effective have attitudes about certain organizational, technical, and behavioral concepts that are different from the attitudes of the less-effective safety directors. Also, the more-effective safety directors tend to have stronger convictions about these concepts than do the less-effective safety directors.
2. Safety directors with nonengineering backgrounds have different attitudes about these concepts than those with engineering backgrounds. Engineers value and have stronger convictions about technical concepts than nonengineers. Also, there is a relationship between the college degree held and managerial effectiveness.

3. Experience affects attitudes toward these concepts and the effectiveness of the safety directors.
4. The organizational position of the safety director affects the amount of power he or she possesses. Also, the amount of power of and the position of the safety director have an influence on his or her effectiveness.

Bracey sent 583 questionnaires (42.5% were returned) to safety directors in four industries: petroleum refineries, chemical plants, electrical manufacturers, and transportation- equipment manufacturers. The questionnaire included a semantic-differential test of seven organizational, technical, and behavioral concepts, and asked questions on organizational characteristics, education, experience of the safety director, and accident statistics. The returns were divided into effective and ineffective classifications based on the accident frequency rates by industry. Bracey found that:

1. The effective safety directors did have different and stronger convictions about the concepts relating to accident prevention.
2. No correlation existed between education and the accident record. A test of attitude differences indicated that education played only a minor role in causing differences in attitudes. It was concluded that engineers and nonengineers make equally good (or bad) safety directors.
3. The portion of the third hypothesis (listed earlier) dealing with experience and attitudes was accepted, while the portion concerning experience and effectiveness was rejected. Experience may be helpful in attaining strong attitudes, but experience does not seem to be related to effectiveness.
4. There was no apparent relationship between the safety director's position in the organizational structure and the amount of power he or she possessed. No correlation was found among position, power, and accident-frequency rates. More powerful safety directors did, however, have stronger convictions about the concepts than the less powerful.

Thus it is apparent that there is no one best kind of safety director, just as there is no one best kind of manager or executive. The best safety manager is the one who manages to assess the situation, the company, and what needs doing, and then gets it done through the people in the organization who must do it.

This suggests that safety-director success is determined by what the line organization accomplishes; that his or her success is dependent on the relationships established with line managers. This is true. Earnest Levens elaborates on some of the problems connected with the relationship:[2]

---

[2]. Reprinted with permission from E. Levens, Some current trends in safety management, *ASSE Journal*, June 1974.

As an example of how inter-personal barriers waste energy, line managers and staff safety professionals often have latent conflicts which they are reluctant to express openly, to the detriment of the loss-prevention program. During the 1970 ASSE Professional Conference, one workshop did some role-playing and listed what bugs managers about safety engineers and vice versa.

These lists are frank and instructive. Think of how much time and effort are dissipated when feelings like these are hidden away while they exert a profound subconscious influence. Think of how much more effectively the loss-prevention effort will be managed in any organization which deals in a real and human way with these very real and very human feelings. We must and can learn to handle change and focus human energy without loss of management responsibility, force, or authority.

The lists mentioned by Levens follow:[3]

What bugs safety engineers about managers? Safety engineers think managers
Overly Cost Conscious
Don't Appreciate Us
Feel They Don't Need Us
Don't Learn Our Point of View
Don't Want to be Told What to Do
Won't Take Free Advice, Just Want Less Work
Duck Off Internal Responsibilities
Overly Concerned with Production at the Expense of People
Not Interested in Their People
No Action Until Someone is Killed
Lip Service, But Not Genuine Support of Safety
Demand no Interference with Production
Make us Responsible for Accidents, But Will Not Give us Authority for Production
Don't Know Safety Program
At Cutback Time, Safety Goes First
We're Too Low in the Hierarchy
Make Safety Cover Up Accidents to Lower Insurance
Don't Believe Good Safety Will Improve Production and Quality

What bugs managers about safety engineers? Managers think safety engineers
Not Time or Cost Conscious
Not Production Oriented

---

3. Ibid.

Can't Measure His Success
Questionable Overhead—Not Worth the Money
Purveyor of Gloom and Doom
No Personal Responsibility for Accidents
More Problems than Solutions
Unprepared, Unobjective, Inconclusive
Nitpicker
Unrealistic, Unreasonable
Authoritarian and Pushes It
Superficial
Doesn't Provide Solutions
Too Theoretical, Often Impractical
Not Productive, Strictly Overhead
Doesn't Understand Management
Hindsight Specialist
Rocks the Boat

The Change Agent and Approaches to Change

The safety professional might best be described as a change agent in the organization. It is his or her role to determine the need for change, to determine how the organization should change, and to make that change happen. Table 19-1 summarizes three general approaches that can be used by the change agent. Almost everything we have said in the last two chapters about change and how to bring it about is relevant here. (See also Figure 19-2.)

In addition to the primary role as change agent, the safety professional might also be responsible for some specific human-error control techniques identified in Chapter 18.

Accident-Prone-Situation Identification and Error-Cause Removal

Much of the accident-prone-situation (APS) identification process, along with error-cause removal (ECR), will probably be a staff-safety-specialist function. If it is not the specialist's direct responsibility, it will surely be his or her function to ensure that line managers are able to handle the techniques.

Table 19-1. Comparison of three general approaches for initiating organizational changes

Approaches for Initiating Change	Typical Interventional Techniques	Intended Immediate Outcomes	Assumptions about the Major Causes of Behavior in Organizations
Individuals	Education, training, socialization, attitude change	Improvements in skill levels, attitudes, and motivation of people	Behavior in organizations is largely determined by the characteristics of the people who compose the organization.
Organizational structure and systems	Modification of actual organizational practices, procedures, and policies which affect what people do at work	Creation of conditions to elicit and reward member behaviors which facilitate organizational goal achievement	Behavior in organizations is largely determined by the characteristics of the organizational situation in which people work.
Organizational climate and interpersonal style	Experiential techniques aimed at increasing members' awareness of the social determinants of their behavior and helping them learn new ways of reacting to and relating to each other within the organizational context	Creation of a system-wide climate which is characterized by high interpersonal trust and openness: reduction of dysfunctional consequences of excessive social conflict and competitiveness	Behavior in organizations is largely determined by the emotional and social processes which characterize the relations among organizational members.

Climate Analysis

This was discussed in some detail in Chapter 5. The analysis suggested that upper management do what the safety professional might prefer to do, or at least be a part of.

Environmental Analysis

This normally refers to industrial hygiene activities, and is fairly traditional.

Human-Factors Analysis

The safety professional might wish to engage in this personally or to obtain outside help.

Directive consultation							
Position 1	Position 2	Position 3	Position 4	Position 5	Position 6	Position 7	Position 8
Advocate	Expert	Trainer	Alternative identifier	Collaborator	Process specialist	Fact-finder	Reflector
Persuades client as to proper approach	Gives expert advice to client	Develops training experiences to aid client	Provides alternative to client	Joins in problem solving	Assists client in problem-solving process	Serves to help client collect data	Serves catalyst age client solv the prob

Non-directive consult

Figure 19-2. Consulting approaches that can be used by the change agent. (From G. Lippitt and Nadler, Emerging roles of the training director. Reproduced by special permission from the August 1967 *Training and Development Journal.* Copyright 1967 by the American Society for Training and Development, Inc.)

Proneness Identification

The identification of accident-prone workers is a valid concept, but the general feeling among safety experts today is that we might get more results concentrating on other areas.

Systems-Safety Check

Systems failure, systems checking, and the whole area of systems safety constitute an area where the safety professional might choose to put his or time and effort.

Attitude-Developer Check

As discussed earlier, attitudes toward safety are developed through the conditions surrounding safety-program instruction, through the consequences of being associated with the safety program, and through modeling. The negatives and positives associated with a subject determine people's feelings about that subject. The attitude-developer check is a systematic way of looking at the work situation to determine what negatives and positives are associated with it. Figure 19-3 is a form that can be used to do this. The safety professional merely observes a job and talks to the worker on that job to see what the worker is experiencing on a regular basis. After a number of these forms have been filled out by the ob-

Role of Staff

	Attitude-Developer Check	
Job	Worker	
List negatives noted on this job Conditions of job		Consequences of work
List positives on this job Conditions of job		Consequences of work
Describe the modeling		
Conclusions/suggestions		

Figure 19-3. Attitudinal measurement form.

server/interviewer, he or she should have a good feel for how and why attitudes of a certain kind are developed.

Boss-Measurement Check

Since motivation is so closely allied to measurement, it seems important to study in some detail the measurement systems in effect in the organization, as read by the worker. Performance is highly influenced by measurement at all levels, particularly at the lower organizational levels. Therefore, it seems important to analyze the measurements. Figure 19-4 can help the evaluator in this task of analysis. Specific workers on specific jobs are contacted for the purpose of determining (1) what, specifically, the job is, and (2) how the worker thinks the boss is measuring his or her safety performance; and, based on this, making (3) an analysis of that measurement as perceived by the person being measured. Is the measure crisp or fuzzy (precise or imprecise)? Is it tightly or loosely connected to the worker's performance? A crisp measure is one based on an accurate analysis of whether or not the person did something. For example, did a worker attend a safety meeting or not? Whether or not the worker has a good safety attitude is a fuzzy measurement. Did the worker complete a job-safety analysis? This is a tight measurement. Whether or not the worker has had an accident is a loose measurement of his or her safety performance.

This boss-measurement check can be used at various levels of the organization. It might be particularly helpful at the supervisory level to determine how supervisors think they are being measured by middle managers.

Boss-Style Check

If the style of management is important to safety results, as we believe it is, then we ought to be examining that variable. Figure 19-5 attempts to provide a simple way of doing this. A number of employees in a department are interviewed, and the leadership style of the boss is assessed. In addition, the boss is observed in

---

Boss-Measurement Check

Job _____ Worker _____

List specific tasks of the job:

What does worker do?	How does the boss know the worker has done it?	Is the measure: Crisp? Fuzzy? Tight? Loose?

Conclusions/suggestions

Figure 19-4. Boss-measurement-check form.

Role of Staff

Boss-Style Check

Department _____ Boss _____

Employee contacted	Assumption		General impression obtained	
	Theory X	Theory Y	Authoritarian	Participative
Your observation				

Conclusions/suggestions

Figure 19-5. Boss-style-check form.

his or her day-to-day leadership of people. The boss's assumptions are assessed (is he or she an X- or Y-theory manager?), and a general impression obtained of his or her style, ranging on a scale from 1 (authoritarian) to 10 (participative).

Priority Analysis

Safety success (and human-error-reduction success) is also dependent on how high a priority safety actually has in the organization in comparison with other management goals. It behooves us to get as accurate a reading of this priority as is possible. Figures 19-6 and 19-7 can help us to get this reading. Interviewing both supervisors and workers and asking them about (1) the time spent on safety

Priority Analysis

Supervisor interviewed _____

Time spent daily: on safety _____   On other matters _____

How does this supervisor think his or her boss is measuring him or her on safety performance?

How does this supervisor think his or her boss "priortizes" safety?

What interaction does this supervisor have with his or her boss relating to safety?

When was it last discussed? _____

What led to the discussion? _____

An accident? _____

An observed unsafe act? _____

An observed unsafe condition? _____

Something an employee did wrong? _____

Something you did wrong? _____

Something the employee did right? _____

Something you did right? _____

Your regular report? _____

What were your feelings about the discussion? _____

Do you feel safety is a key goal of your boss? Why? _____
Have the supervisor rank these:
Production
Quality control
Cost control
Customer service
Safety
Equal opportunity for all

Conclusions/suggestions

Figure 19-6. Priority-analysis form to be used with supervisors.

Priority Analysis

Employee interviewed _____

How does this employee think his or her boss is measuring him on safety?

How does this employee think his or her boss prioritizes safety?

What interaction does this employee have with his or her boss relating to safety?

When was it last discussed? _____

What led to the discussion? _____

An accident? _____

An unsafe act he or she saw? _____

An unsafe condition he or she spotted? _____

Something you did wrong? _____

Something you did right? _____

What were your feelings about the discussion? _____

How important do you feel your boss thinks safety is? _____

How would the employee rank these?
Production
Quality control
Cost control
Customer service safety
Equal opportunity

Conclusions/suggestions

Figure 19-7. Priority-analysis form to be used with employees.

Group-Analysis Check
Department _____ Group _____
Did you observe these characteristics of a weak group?
Were there cliques and subgroups? _____
Was cooperation poor in the group? _____
Were they unfriendly to one another? _____
Was initiative lacking? _____
Do they avoid responsibility? _____
Is there little respect shown for the company? _____
Did you observe these characteristics of a strong group?
Do they try to deserve group praise? _____
Do they seek group recognition? _____
Do they put pressure on one another? _____
Is there special effort for group goals? _____
Is there individual competence? _____
Is there maturity in the members? _____
Is there individual strength? _____
Are there common goals? _____
Are they in physical proximity? _____
Who are their leaders? _____
Do they use any group symbols? _____
Who/what do they compete against? _____
Do they seem to like one another? _____
How do they feel about safety/safety rules?
Rules they buy:
Rules they reject:
Conclusions/suggestions

Figure 19-8. Group-analysis-check form.

in relation to time spent on other tasks, (2) how they think they are measured, (3) what they think their bosses' priorities are, and (4) how often contacts are made in the name of safety, should help us to figure out what management's actual priority system is.

Group-Analysis Check

Psychologists and sociologists state that the peer group is the largest single determinant of employee behavior on the job. If that is true, and it seems to be, we ought to be examining just how influential it is, and in what direction it pulls the employee. Figure 19-8 helps us to look at the informal group, assess its strengths and weaknesses, and discover whether or not it seems to be for or against safety.

REFERENCES

American Society of Safety Engineers. The scope and functions of the safety professional. Park Ridge, IL: The Society, 1966.

Bracey, H. An investigation into the effectiveness of safety directors as influenced by selected variables. Photocopied. Baton Rouge: Louisiana State University, 1969.

Levens, E. Some current trends in safety management. *ASSE Journal,* 1974.

Lippitt, G. and L. Nadler. Emerging roles of the training director. *Training and Development Journal,* August 1967.

PART *VI* | *Appendixes*

Appendix A

Stress Tests (Individuals)

First, what are the warning signs of stress in an individual? Here are some:

- General irritability, hyperexcitation, or depression
- Pounding of the heart
- Dryness of throat and mouth
- Impulsive behavior, emotional instability
- Overpowering urge to cry or run and hide
- Inability to concentrate
- Feelings of unreality, weakness, dizziness
- Predilection to becoming fatigued
- Floating anxiety
- Tension and alertness
- Trembling or nervous ticks
- Tendency to be easily startled
- High-pitched nervous laughter
- Stuttering or other speech difficulties
- Bruxism (grinding teeth together)
- Insomnia
- Hypermotility, moving for no reason, inability to relax
- Sweating
- Frequent need to urinate
- Diarrhea, indigestion, queasiness, vomiting
- Migraine headaches
- Premenstrual tension or missed cycles
- Pain in neck and lower back
- Loss of or excessive appetite

- Increased smoking
- Increased use of drugs or alcohol
- Drug or alcohol addiction
- Nightmares
- Neurotic behavior
- Psychosis
- Accident proneness

INDIVIDUAL TESTS

Are you in one of these categories?

1. Laborer
2. Secretary
3. Inspector
4. Lab technician
5. Office manager
6. Foreman
7. Manager
8. Waitress
9. Machine operator
10. Farm owner
11. Mine operator
12. Printer

How many of the following are you experiencing?

1. Job satisfaction
2. Physical conditions
3. Organizational factors
4. Work load
5. Work hours
6. Work task
7. Career development
8. Downsizing
9. Big Brother syndrome
10. Takeovers
11. Too much responsibility
12. The ostrich syndrome

TESTING FOR PROPENSITY

In recent years a number of tests have been developed to help assess the individual's propensity to stress problems. (See Figures A-1 through A-9.) Each is presented with a brief description of what the test intends to do and how to score it. Later in the chapter is a summary sheet for all scores.

Each test is a self-assessment device, each self-scored. The reader is urged to honestly take each and record scores on the summary sheet, which, when com-

Stress Tests (Individuals)

Below are listed events which occur in the process of living. Place a check in the left-hand column for each of those events that have happened to you during the last 12 months.

Life Event	Point Values
Death of spouse	100
Divorce	73
Marital separation	65
Jail term	63
Death of close family member	63
Personal injury or illness	53
Marriage	50
Fired from work	47
Marital reconciliation	45
Retirement	45
Change in family member's health	44
Pregnancy	40
Sex difficulties	39
Addition to family	39
Business readjustment	39
Change in financial status	38
Death of close friend	37
Change to different line of work	36
Change in number of marital arguments	35
Mortgage or loan over $60,000	31
Foreclosure of mortgage or loan	30
Change in work responsibilities	29
Son or daughter leaving home	29
Trouble with in-laws	29
Outstanding personal achievement	28
Spouse begins or stops work	26
Starting or finishing school	26
Change in living conditions	25
Revision of personal habits	24
Trouble with boss	23
Change in work hours, conditions	20
Change in residence	20
Change in schools	20
Change in recreational habits	19
Change in church activities	19
Change in social activities	18
Mortgage or loan under $60,000	17
Change in sleeping habits	16
Change in number of family gatherings	15
Change in eating habits	15
Vacation	13
Christmas season	12
Minor violations of the law	11

Score: _____

After checking the items above, add up the point values for all of the items checked.

Figure A-1. Self-assessment Exercise 1 on need to adapt.

Choose the most appropriate answer for each of the 10 statements below as it usually pertains to you. Place the letter of your response in the space to the left of the question.

___ 1. When I can't do something "my way," I simply adjust to do it the easiest way.
  (a) Almost always true (b) Often true
  (c) Seldom true (d) Almost never true

___ 2. I get "upset" when someone in front of me drives slowly.
  (a) Almost always true (b) Often true
  (c) Seldom true (d) Almost never true

___ 3. It bothers me when my plans are dependent upon the actions of others.
  (a) Almost always true (b) Often true
  (c) Seldom true (d) Almost never true

___ 4. Whenever possible, I tend to avoid large crowds.
  (a) Almost always true (b) Often true
  (c) Seldom true (d) Almost never true

___ 5. I am uncomfortable having to stand in long lines.
  (a) Almost always true (b) Often true
  (c) Seldom true (d) Almost never true

___ 6. Arguments upset me.
  (a) Almost always true (b) Often true
  (c) Seldom true (d) Almost never true

___ 7. When my plans don't "flow smoothly," I become anxious.
  (a) Almost always true (b) Often true
  (c) Seldom true (d) Almost never true

___ 8. I require a lot of room (space) to live and work in.
  (a) Almost always true (b) Often true
  (c) Seldom true (d) Almost never true

___ 9. When I am busy at some task, I hate to be disturbed.
  (a) Almost always true (b) Often true
  (c) Seldom true (d) Almost never true

___ 10. I believe that "all good things are worth waiting for."
  (a) Almost always true (c) Often true
  (c) Seldom true (d) Almost never true

Scoring: 1 and 10: a = 1, b = 2, c = 3, d = 4   Score:_____
  2–9: a = 4, b = 3, c = 2, d = 1

Figure A-2. Self-assessment Exercise 2 on ability to cope with frustration.

Choose the most appropriate answer for each of the 10 statements below and place the letter of your response in the space to the left of the question.

How often do you:
___ 1. Feel like you have to work overtime to complete your work?
    (a) Almost always   (b) Very often
    (c) Seldom   (d) Never
___ 2. Feel like your boss (the organization) forces you to work overtime?
    (a) Almost always   (b) Very often
    (c) Seldom   (d) Never
___ 3. Feel like you need an assistant?
    (a) Almost always   (b) Very often
    (c) Seldom   (d) Never
___ 4. Feel confused as to exactly what your role is?
    (a) Almost always   (b) Very often
    (c) Seldom   (d) Never
___ 5. Feel discouraged because you can't get to everything that needs to be done?
    (a) Almost always   (b) Very often
    (c) Seldom   (d) Never
___ 6. Feel depressed because there isn't time to do quality work?
    (a) Almost always   (b) Very often
    (c) Seldom   (d) Never
___ 7. Feel the need to skip lunch or coffee breaks to finish tasks?
    (a) Almost always   (b) Very often
    (c) Seldom   (d) Never
___ 8. Believe you do not have enough authority?
    (a) Almost always   (b) Very often
    (c) Seldom   (d) Never
___ 9. Are being judged by things beyond your control?
    (a) Almost always   (b) Very often
    (c) Seldom   (d) Never
___ 10. Feel that work that must be done at home interferes with the time needed at work?
    (a) Almost always   (b) Very often
    (c) Seldom   (d) Never

Scoring: a = 4, b = 3, c = 2, d = 1   Score: _____

Figure A-3. Self-assessment Exercise 3 on perception of overload.

Indicate the most appropriate answer for each of the 10 statements in the space provided.

___ 1. I need more social situations.
    (a) Almost always true   (b) Often true
    (c) Seldom true   (d) Almost never true

___ 2. At times I feel very much alone.
    (a) Almost always true   (b) Often true
    (c) Seldom true   (d) Almost never true

___ 3. I dislike travelling and would rather be home with my family.
    (a) Almost always true   (b) Often true
    (c) Seldom true   (d) Almost never true

___ 4. Whenever I have a free minute I find something to read.
    (a) Almost always true   (b) Often true
    (c) Seldom true   (d) Almost never true

___ 5. I find a job transfer and move very difficult.
    (a) Almost always true   (b) Often true
    (c) Seldom true   (d) Almost never true

___ 6. I need a great deal of variety in my work; I get bored easily.
    (a) Almost always true   (b) Often true
    (c) Seldom true   (d) Almost never true

___ 7. I detest standing in lines and waiting.
    (a) Almost always true   (b) Often true
    (c) Seldom true   (d) Almost never true

___ 8. Listening to a lecture usually bores me.
    (a) Almost always true   (b) Often true
    (c) Seldom true   (d) Almost never true

___ 9. I'm a pack rat and find it difficult to throw things away.
    (a) Almost always true   (b) Often true
    (c) Seldom true   (d) Almost never true

___ 10. I dislike not being with others.
    (a) Almost always true   (b) Often true
    (c) Seldom true   (d) Almost never true

Scoring: a = 4, b = 3, c = 2, d = 1   Score: _____

Figure A-4. Self-assessment Exercise 4 on perception of deprivation.

Stress Tests (Individuals)

Answer the following questions; place the letter of your response in the space to the left.

___ 1. How many cigarettes do you smoke daily?
    (a) None   (b) Less than 1 pack
    (c) 1–2 packs   (d) More than 2 packs

___ 2. How many alcoholic drinks do you average in a day.
    (a) None   (b) 1
    (c) 1–2   (d) More than 2

___ 3. How many cups of coffee do you drink each day?
    (a) 2 or less   (b) 3–4
    (c) 5–6   (d) 7 or more

___ 4. How much exercise do you get daily?
    (a) More than 30 minutes (b) Less than 15 minutes
    (c) 15–30 minutes   (d) None

___ 5. How often do you snack between meals?
    (a) Never   (b) Once a week
    (c) Once a day   (d) Between each meal

___ 6. How many of the following have you been treated for:
    ulcers   migraines   hypertension   hives   allergies   black-outs
    (a) None   (b) One
    (c) Two   (d) More than two

___ 7. How many of the following do you practice regularly?
    transcendental mediation   relaxation response   other meditation
    (a) Three   (b) Two
    (c) One   (d) None

___ 8. Are you overweight?
    (a) No   (b) 5–10 pounds
    (c) 10–20 pounds   (d) More than 20 pounds

___ 9. Do you take vitamins daily?
    (a) Yes   (b) Yes
    (c) No   (d) No

___ 10. How often do you travel on business?
    (a) Never   (b) 1–2 trips per year
    (c) 1 trip per month   (d) More than 1 trip per month

Scoring: a = 1, b = 2, c = 3, d = 4   Score: _____

Figure A-5. Self-assessment Exercise 5 on health habits.

Calculate your exposure to noise using the following chart.

Example	Overall sound pressure level (dB re 0.0002 microbar)	Hours exposed daily	
	10×	_____	= ____
Studio for sound pictures	20×	_____	= ____
Soft whisper (5 feet)	30×	_____	= ____
Quiet office	40×	_____	= ____
Audiometric testing booth	×	_____	= ____
Average residence	50×	_____	= ____
Large office	×	_____	= ____
Conversational speech (3 feet)	60×	_____	= ____
Freight train (100 feet)	70×	_____	= ____
Average automobile (30 feet)	74×	_____	= ____
Very noisy restaurant	80×	_____	= ____
Average factory	×	_____	= ____
Subway	90×	_____	= ____
Printing press plant	×	_____	= ____
Looms in textile mill	100×	_____	= ____
Electric furnace area	×	_____	= ____
Woodworking	110×	_____	= ____
Casting shakeout area	×	_____	= ____
Hydraulic press	120×	_____	= ____
50 hp siren (100 feet)	×	_____	= ____
Jet plane	140×	_____	= ____
Rocket launching pad	180×	_____	= ____

Total noise level = _____

Divide by 8 = _____
(Average hourly level)

Figure A-6. Self-assessment Exercise 6 to measure exposure to noise.

Choose the alternative that best summarizes how you generally behave, and place your answer in the space provided.

___ 1. When I face a difficult task, I try my best and will usually succeed.
    (a) Almost always true   (b) Often true
    (c) Seldom true   (d) Almost never true

___ 2. I am at ease when around members of the opposite sex.
    (a) Almost always true   (b) Often true
    (c) Seldom true   (d) Almost never true

___ 3. I feel that I have a lot going for me.
    (a) Almost always true   (b) Often true
    (c) Seldom true   (d) Almost never true

___ 4. I have a very high degree of confidence in my own abilities.
    (a) Almost always true   (b) Often true
    (c) Seldom true   (d) Almost never true

___ 5. I prefer to be in control of my own life as opposed to having someone else make decisions for me.
    (a) Almost always true   (b) Often true
    (c) Seldom true   (d) Almost never true

___ 6. I am comfortable and at ease around my superiors.
    (a) Almost always true   (b) Often true
    (c) Seldom true   (d) Almost never true

___ 7. I am often overly self-conscious or shy when among strangers.
    (a) Almost always true   (b) Often true
    (c) Seldom true   (d) Almost never true

___ 8. Whenever something goes wrong, I tend to blame myself.
    (a) Almost always true   (b) Often true
    (c) Seldom true   (d) Almost never true

___ 9. When I don't succeed, I tend to let it depress me more than I should.
    (a) Almost always true   (b) Often true
    (c) Seldom true   (d) Almost never true

___ 10. I often feel that I am beyond helping.
    (a) Almost always true   (b) Often true
    (c) Seldom true   (d) Almost never true

Scoring: 1–6: a = 1, b = 2, c = 3, d = 4   Score:_____
          7–10: a = 4, b = 3, c = 2, d = 1

Figure A-7. Self-assessment Exercise 7 on perception of self.

Answer the following questions; place the letter of your response in the space to the left.

Do you:

___ 1. Often have tense muscles?
    (a) Almost always true   (b) Often true
    (c) Seldom true   (d) Almost never true

___ 2. Often experience dryness in the mouth?
    (a) Almost always true   (b) Often true
    (c) Seldom true   (d) Almost never true

___ 3. Sweat profusely?
    (a) Almost always true   (b) Often true
    (c) Seldom true   (d) Almost never true

___ 4. Have difficulty expressing yourself?
    (a) Almost always true   (b) Often true
    (c) Seldom true   (d) Almost never true

___ 5. Try to solve all problems immediately?
    (a) Almost always true   (b) Often true
    (c) Seldom true   (d) Almost never true

___ 6. Create crises when there are none?
    (a) Almost always true   (b) Often true
    (c) Seldom true   (d) Almost never true

___ 7. Dwell on past crises?
    (a) Almost always true   (b) Often true
    (c) Seldom true   (d) Almost never true

___ 8. Experience hand tremors?
    (a) Almost always true   (b) Often true
    (c) Seldom true   (d) Almost never true

___ 9. Have a pounding heart occasionally?
    (a) Almost always true   (b) Often true
    (c) Seldom true   (d) Almost never true

___ 10. Experience hives?
    (a) Almost always true   (b) Often true
    (c) Seldom true   (d) Almost never true

Scoring: a = 4, b = 3, c = 2, d = 1   Score:_____

Figure A-8. Self-assessment Exercise 8 to measure anxious reactivity.

```
Answer the following questions.

Do you:                                                Yes    No

 1. Eat, talk and walk quickly?                        ___   ___
 2. Get easily bored? Tune people out?                 ___   ___
 3. Get impatient with slow people?                    ___   ___
 4. Feel guilty when relaxing?                         ___   ___
 5. Forget small details?                              ___   ___
 6. Usually speak rapidly?                             ___   ___
 7. Like to own things?                                ___   ___
 8. Generally lean forward on your chair?              ___   ___
 9. Measure yourself by goals achieved?                ___   ___
10. Do everything at a rapid pace?                     ___   ___


       Score: yes = 4, no = 1                   Score: ___
```

Figure A-9. Self-assessment Exercise 9 to measure Type A behavior.

Exercise:	1	2	3	4	5	6	7	8	9
	Adaptation	Frustration	Overload	Deprivation	Health	Noise	Self-Perception	Anxious Reactivity	Type A Behavior
	• 400	• 40	• 40	• 40	• 40	•	• 40	• 40	• 40
	•	•	•	•	•	•	•	•	•
	•	•	•	•	•	•	•	•	•
	•	•	•	•	•	•	•	•	•
Scores Indicative of High Vulnerability to Stressors	• 350	• 35	• 35	• 35	• 35	• 105	• 35	• 35	• 35
	•	•	•	•	•	•	•	•	•
	•	•	•	•	•	•	•	•	•
	•	•	•	•	•	•	•	•	•
	• 300	• 30	• 30	• 30	• 30	• 95	• 30	• 30	• 30
	•	•	•	•	•	•	•	•	•
	•	•	•	•	•	•	•	•	•
	•	•	•	•	•	•	•	•	•
Moderate Vulnerability to Stressors	• 250	• 25	• 25	• 25	• 25	• 85	• 25	• 25	• 25
	•	•	•	•	•	•	•	•	•
	•	•	•	•	•	•	•	•	•
	•	•	•	•	•	•	•	•	•
	• 200	• 20	• 20	• 20	• 20	• 75	• 20	• 20	• 20
	•	•	•	•	•	•	•	•	•
	•	•	•	•	•	•	•	•	•
Low Vulnerability to Stressors	• 150	• 15	• 15	• 15	• 15	• 65	• 15	• 15	• 15
	•	•	•	•	•	•	•	•	•
	•	•	•	•	•	•	•	•	•
	•	•	•	•	•	•	•	•	•
	• 100	• 10	• 10	• 10	• 10	•	• 10	• 10	• 10

Figure A-10. Personal stress profile summary sheet.

pleted, will give a picture of the total propensity profile, showing the current situation and the areas in which needed improvement might be made.

These nine self-assessment exercises (Figures A-1 through A-9) provide a beginning point in assessing an individual's (or your) propensity to stress problems.

- *Exercise 1,* which was described earlier, is repeated here for convenience. It looks at the need to adapt.
- *Exercise 2* looks at behavior when faced with frustrations and at the ability to cope in these situations.
- *Exercise 3* measures the perception of overload.
- *Exercise 4* measures the perception of deprivation.
- *Exercise 5* measures health habits.
- *Exercise 6* looks at exposure to noise.
- *Exercise 7* measures perception of self.
- *Exercise 8* measures anxious reactivity.
- *Exercise 9* measures Type A behavior (discussed earlier).

The nine exercises provide a profile of your propensity to stress. Figure A-10 is the summary sheet for all the scores.

Appendix B

CTD Analysis

An ergonomic analysis is a way of reducing the probability that workers will be injured through exposure to cumulative trauma in repetitive motion. Each job they do should be analyzed.

This ergonomic analysis has as its primary focus the prevention of repetitive motion injuries, which are commonly called cumulative trauma disorders (CTDs). These CTDs are occupational injuries that develop over time in any part of the body, but they are most prevalent in the arms and the back. These injuries are caused by jobs that require repeated exertions and movements near the limits of the individual's strength and range-of-motion capability. If the individual continues the action that is causing the pain, cumulative trauma disorders are likely to develop.

CTDs can have the following effects on individuals:

- Pain
- Numbness or loss of sensation
- Reduced strength
- Degraded ability to perform work

Cumulative trauma disorders have been correlated with hazardous combinations involving the following:

- Forceful exertions
- Frequent exertions
- Body posture
- Mechanical stress
- Vibration

APPENDIXES

Hand & Wrist	Possible CTD Problem
Grasp • pinch grip • static hold	• prolonged pinch grip • forceful grip
Wrist Posture • flexion/extension • ulnar/radial deviation	• flex/ext. >45° • radial/ulnar deviation
Frequency • hand or wrist manipulations	• >10 per minute
Mechanical Stress • localized pressure to palm or fingers • scraping/bumping • strike with hand • single finger trigger	• prolonged exposure • intense exposure
Vibration • high frequency vibration	• prolonged exposure

Neck	Possible CTD Problem
Neck Posture • bend/twist >20°	• >50% of time
Back	
Lifting	• >90% of AL • non-NIOSH
Torso Posture • torso bending • torso twisting	• bend >45° • bend >20° + twist
Static Hold/Carry • >5 seconds	• >10 lbs. • 5–10 lbs. with flexed shoulder
Static Load • not able to change sit/stand posture over work day	• poor posture
Push/Pull • whole body action	• poor conditions

Arm & Shoulder	Possible CTD Problem
Arm Work • exertions > 5 lbs.	• little rest between exertions • large exertion
Static Load • prolonged holding	• unsupported • large exertion
Elbow Posture • fully flexed • fully extended • rotated forearm	• repeatedly
Shoulder Posture • flexed • extended • abducted	• flex/abduct > 90° • any extension
Mechanical Stress • sharp edges • hard surfaces	• repeatedly • high force

Legs	Possible CTD Problem
Foot Actuation • foot pedals	• excessive force • extreme posture • high frequency or duration
Leg Posture • knee • ankle	• deep squat • kneeling • 1 legged posture • walk/stand on uneven surface
Mechanical Stress • localized pressure • kicking	• high force • prolonged exposure

Figure B-1. Relationship of CTD problems to body parts.

CTD Analysis

Company _____	CTD Problem?	
Dept. _____	**LEFT HAND & WRIST**	
Supervisor _____	Grasp	
Job Name _____	Wrist Posture	
	Frequency	
	Mechanical Stress	
	Vibration	
	RIGHT HAND & WRIST	
	Grasp	
	Wrist Posture	
	Frequency	
Tools/Parts Weight	Mechanical Stress	
	Vibration	
	ARM & SHOULDERS	
	Arm Work	
	Static Load	
	Elbow Posture	
	Shoulder Posture	
COMMENTS	Mechanical Stress	
	NECK	
	Neck Posture	
	BACK	
	Lifting	
	Torso Posture	
	Static Hold/Carry	
	Static Load	
	Push/Pull	
	LEGS	
	Foot Actuation	
	Leg Posture	
	Mechanical stress	

Figure B-2. CTD analysis form.

Figure B-1 shows for each individual part of the body each of the factors that must be looked at in an ergonomic analysis, and at what point you should begin to be concerned that a factor could eventually cause a CTD.

Using the form in Figure B-2, each job can be analyzed.

Appendix C

Ergonomic Data

In this appendix background information is presented on some basic ergonomic data that can be used for design and ergonomic analysis purposes. Most of the information comes from *Human Engineering Guide to Equipment Design,* by Van Cott and Kinkade, published by the U.S. Government Printing Office.

ANTHROPOMETRIC CONSIDERATIONS

Anthropometric considerations involve the relationships among body dimensions, strength, dexterity, and mobility of the product user. It is easy to understand that because of differences between large people and small people, strong people and weak people, it may be difficult to design and fit everyone, and in some cases economically unfeasible.

Generally speaking, one should consider larger people where clearance is concerned, small people where reach is concerned, and weak people when force is concerned. Mobility and dexterity are not so much a matter of size. However, there is an interaction which must be considered. For example, if a product has to be used in a position that is selected as a compromise for both ends of the size population, one or the other (i.e., large or small) may be forced to assume positions that interfere with mobility and dexterity.

Basic Body Dimensions

- Stature—This dimension should be considered wherever head clearance may be a problem. Since an overhead that clears the tallest person also accommodates the shortest person, this value is the determining criterion.

- Eye height—The position of the operator's eyes is important in determining proper placement of visual displays and for providing adequate clearance for seeing over intervening objects.
- Shoulder height—Generally, it is recommended that controls be placed somewhere between waist and shoulder height for most convenient operation.
- Arm reach—The person with the shortest arm reach generally establishes the primary constraint in deciding where to locate controls.

Normal Habit Patterns

A number of useful and operationally meaningful conclusions can be drawn about human behavior. These observations represent simplifications and generalizations about some very complex, normal human behavior.

Table C-1 contains some of the recognized observations about typical human behavior characteristics that may lead to injury or to unsafe acts. Each statement is followed by a design-oriented precaution or injunction, the summarized results of which are subsequently incorporated in human engineering safe-design criteria listings.

Population Stereotypes

The term population stereotypes has more than one meaning. The most common meaning relates to what people do to or with things and the extent to which this behavior becomes relatively universal across both geography and time. For example, a very common behavioral stereotype is for persons to *push* outward on exit doors and to *pull* towards themselves on entry doors to public buildings.

Common practice in the United States is for electric switches to be pushed up or to the right for ON; whereas for fluid control, the value is turned counter-clockwise for increase or ON and clockwise for decrease or OFF.

In Europe, and much of the rest of the world, these conventions do not apply; thus the specific population stereotype may be the opposite of that which prevails in the United States.

Ergonomic Data

Table C-1. List of typical human behaviors that may lead to an unsafe act or to injury

Behavior Description	Design Consideration
Many people do not consider the effects of surface friction on their ability to grasp and hold an article.	Design surface texture to provide friction characteristics commensurate with functional requirements of task or device.
Most people cannot estimate distances, clearances, or velocities very well; people tend to over-estimate short distances and underestimate long distances.	Design products so that users need not make estimates of critical distances, clearances, or speeds. Provide indicators of these quantities where there is a functional requirement.
Most people do not look where they put their hands and feet, especially in familiar surroundings.	If hand or foot placement is a critical aspect of the user–product interface, design so that careless, inadvertent placement of hands or feet will not result in injury. Provide guards, restraints, and warning labels.
People often utilize the first thing handy as an aid in getting to where they want to go or to manipulate something.	Either design the product so that the "first thing handy" simply cannot be functionally useful, or so that it can!
Many people do not think about such things as: high temperatures; becoming an electrical ground when they grasp an article which is "hot"; picking up an object that is wet; or walking carefully on a slick floor.	This lack of awareness may result from ignorance, inattention, carelessness or sheer disregard for safety. Where possible, observe "Murphy's Law"—design so the person simply cannot make the wrong motion, or touch the "hot" spot.
People seldom anticipate the possibility of contact with sharp edges of corners.	Except to meet functional requirements, eliminate sharp edges on surfaces or units where inadvertent human contact is even remotely possible.
People rarely think about the possibility of catching their clothing on a handle or other protruding object.	Same as preceding item; also provide proper warnings and labels. Conduct user task (operator function) analysis, and always mount handles in most functionally advantageous places.
Very few people think about the possibility of fire or explosion from overheated objects or cooking oil suddenly exposed to the air.	Unless it is a functional requirement, eliminate configurations which will permit such possibilities, even with product misuse.
People seldom give serious thought to the innate curiosity of children in exploring the wonderful mysteries of appliances, pots, and pans.	Obviously, no simple injunction approaches adequacy; follow all available human engineering specifications and safety precautions. Eliminate sharp edges, i.e., heavy spring-loaded closures. Use glass and ceramic material with special fracture characteristics.
Many people do not take the time to read labels or instructions, or to observe safety precautions.	Make labels *brief, bold, simple, clear.* Repeat (place) same labels on various parts of a product. Make use of color coding, fail-safe innovations, and other attention-demanding devices.

Table C-1. (continued)

Behavior Description	Design Consideration
Many designers do not recognize the existence of response stereotypes, i.e., that the average user "expects" something to operate in a certain fashion.	Become aware of the more common stereotype behaviors (see material in this appendix). Also, do not depart from common design on objects or items, which have demonstrated utility and user acceptance just for sake of change, i.e., standard positions for ON in wall electric switch is *UP;* don't change to *DOWN.*
Most people have little mechanical aptitude, and, therefore, do not recognize mechanical relationships which would suggest a mode of operation.	For most home-use products, routine operation and handling *can* be made very simple. For example, even in such complex items as cassette-type sound recorders and play-back units there is a tremendous range in operational use-complexity, all for accomplishing essentially the same function. Keep it simple.
Many people have a penchant for operating an appliance while their hands are full.	If possible, design so that they can do just this! Conversely, design so that the hand/product interface is a full challenge to the operator.
Many people perform most tasks while thinking about something else.	Accept this as a way of life; design so they can continue this practice. No really satisfactory design-out on this one, except— KEEP IT SIMPLE.
Most people perform in a perfunctory manner, utilizing previous habit patterns. Under stress, they almost always revert to these habit patterns.	Don't change an established design (if satisfactory) just for the sake of change. Base "design innovations" on functional requirements changes.
Most people will use their hands to test, examine, or explore.	Design in such a way that exploring hands and feet won't get hurt.
Many people will continue to use a faulty article even though they suspect that it may be dangerous.	Two general alternatives: (1) do not build in "graceful degradation" characteristics; or (2) provide clearly marked, evident, fault or failure warning indicators.
Very few people recognize the fact that they cannot see well enough either because of poor eyesight or because of lack of illumination.	Make all critical labels such that they can be clearly and easily read by individuals with no better than 20–40 vision (uncorrected) in one eye. Provide task, item, or area illumination within product where functional requirements justify same.
Children act on impulse even though they have been told *to* do or *not* to do something.	Any designer who has children is aware of this behavior tendency. Task or operator requirement analysis may lead to specific solutions.

Ergonomic Data   *343*

Table C-1.   (continued)

Behavior Description	Design Consideration
Most people are reluctant to (and seldom do) recheck their operational or maintenance procedures for errors or omissions.	Keep it simple; provide concise, brief instructions; design for step-by-step procedures with a minimum of procedural interactions and interdependencies. For example, during recent product survey (riding lawn mowers) two very similar units were compared for simplicity of operation; both performed identical functions and outward appearances were strikingly similar. However, one machine had such complicated instructions and such a confusion of levers and controls that it almost defied ordinary usage or understanding.
In emergency situations, people very often respond irrationally and with seemingly random behavior patterns.	Same as preceding item and many others. Keep operation simple; provide for fail-safe operation; follow stereotypes and standard configuration precepts where possible.
People often are unwilling to admit errors or mistakes of judgment or perception, and thus will continue a behavior or action originally initiated in error.	Keep your design simple; provide for fail-safe operation. Where functionally justifiable, design for sequence-checking. Design for automatic product shut-off in the event appropriate sequence is not followed.
Foolish attitudes and emotional biases often force people into apparently irrational behaviors and improper use of products.	Be aware of this. No matter how you design it, there is one end of a gun you can't make safe *and* simultaneously meet the functional requirements of guns.
A physically handicapped person will often undertake tasks and operations of which he or she is incapable, largely out of false pride.	Study the limitations of the physically handicapped. Design for one-hand, one-eye, one-leg operators.
People often misread or fail to see labels, instructions, and scale markers on various items, thereby improperly setting or adjusting them and thus creating a hazard.	Follow detailed human engineering practice and techniques for placement of labels, design of scales, displays and markers.
Many people become complacent after long-term successful use of (or exposure to) generally hazardous products.	Design in such a way that the fail-safe characteristics of a device cannot be avoided by simple means. Provide attention-demanding but simple, brief warning displays. Consider using very dramatic warning devices to indicate potential failure such as flashing lights or loud sounds. In some cases provide for fairly complicated operating procedures associated with automatic system shutdown where sequence of steps is violated.

Visual Displays Design

Although engineering-oriented persons learn from experience to interpret rather complex displays, the majority of product users do not have this experience. Therefore, displays should be selected which are simple to interpret. Several factors should be considered in selecting or designing a visual display. For example, the display should not contain any more information than absolutely necessary to convey the intended message.

In addition, display formats should follow certain rules based upon the typical expectations of the average operator. For example, people interpret or read out displayed information according to certain stereotypes (e.g., they read from left to right, from top to bottom; they expect values to increase from bottom to top or left to right, or clockwise).

Other display formatting rules include sizing and spacing of display. For example, the position of a switch provides system condition information; thus the switch position and/or direction of movement should be compatible with the same rules noted above.

Instructions and Labeling

It should be assumed that all products may at some time be used or operated by someone who is unfamiliar with the product and inexperienced in its use. It should not be assumed that instructions provided by means of an instruction book, pamphlet, or sheet will be used or retained for later reference. Clear, visible, legible instruction labels should be placed on the product in a conspicuous place. All controls and displays, as well as critical parts, also should be labeled. Do not assume that the user will recognize a component or understand what it is for, and how it is to be used.

Control Mode and Design

- To provide an increase in function, control motion should be up, forward, to the right, or clockwise.
- Go/No-Go or ON–OFF controls should operate so that the GO or ON position is up, forward, to the right.
- Push–pull handles should pull out of a control panel for ON.

Ergonomic Data

- Handles used for braking action should "pull" ON.
- Finger-actuated "trigger-type" controls should be compatible with a squeeze action of the operator's hands and fingers; i.e., contraction of finger actuates function.
- A control lever path should be designed so that it is normal to the pivot point of the limb (i.e., radiates out from the point rather than perpendicular to it).
- All foot-operated controls should be pushed for ON or increase in function.
- All controls shall be located within maximum reach limits of the operator when he or she is in the normal operating position. Controls, preferably, should be located at less than 75% of the maximum reach. Controls should be located where they can be reached by a direct or straight path. Panel controls should never be located behind other controls (behind a steering wheel, for example).
- Controls should not be located where they cannot be seen.
- The most frequently used controls should be located in the most convenient place, near the hand or foot normally used to operate them.
- When a control may be used from one or more operating position, it should be located for the best compromise between the two positions.
- Avoid locating controls below waist or above shoulder height. If short, fast reaction time is important, locate controls directly in front of the operator, not on the side.
- Provide sufficient spacing between controls to preclude inadvertent actuation. When there is not enough space, provide separation guards or use a dual-mode safety design.

ERGONOMIC TABLES

Tables C-2 through C-9 and Figures C-1 through C-3 are helpful for design or for making the ergonomic analysis.

Visual Recommendations

Tables C-2 through C-7 and Figure C-1 involve visual considerations.

Table C-2. General illumination levels and types of illumination for different task conditions and types of tasks

Task condition	Type of task or area	Illuminance level (Ft.-c)	Type of illumination
Small detail, low contrast, prolonged periods, high speed, extreme accuracy.	Sewing, inspecting dark materials, etc.	100	General plus supplementary, e.g., desk lamp.
Small detail, fair contrast, speed not essential.	Machining, detail drafting, watch repairing, inspecting medium materials, etc.	50–100	General plus supplementary.
Normal detail, prolonged periods.	Reading, parts assembly; general office and laboratory work.	20–50	General, e.g., overhead ceiling fixture.
Normal detail, no prolonged periods.	Washrooms, power plants, waiting rooms, kitchens.	10–20	General, e.g., random natural or artificial light.
Good contrast, fairly large objects.	Recreational facilities.	5–10	General.
Large objects.	Restaurants, stairways, bulk-supply warehouses.	2–5	General.

Table C-3. Approximate reflectance factors for various surface colors

Color	Reflectance	Color	Reflectance
White	85		
Light:		Dark:	
Cream	75	Gray	30
Gray	75	Red	13
Yellow	75	Brown	10
Buff	70	Blue	8
Green	65	Green	7
Blue	55		
Medium:		Wood Finish:	
Yellow	65	Maple	42
Buff	63	Satinwood	34
Gray	55	English Oak	17
Green	52	Walnut	16
Blue	35	Mahogany	12

Table C-4. Recommendations for indicator, panel, and chart lighting

Condition of use	Lighting technique	Luminance of markings (ft.-L)	Brightness adjustment
Indicator reading, dark adaption necessary.	Red flood, integral or both, with operator choice.	0.02–0.1	Continuous throughout range.
Indicator reading, dark adaptation not necessary but desirable.	Red or low-color-temperature white flood, integral, or both, with operator choice.	0.02–1.0	Continuous throughout range.
Indicator reading, dark adaptation not necessary.	White flood...	1–20...	Fixed or continuous.
Reading of legends on control consoles, dark adaptation necessary.	Red integral lighting red flood, or both, with operator choice.	0.02–0.1...	Continuous throughout range.
Reading of legends on control consoles, dark adaptation not necessary.	White flood...	1–20...	Fixed or continuous.
Possible exposure to bright flashes.	White flood...	10–20	Fixed.
Very high altitude, daylight restricted by cockpit design.	White flood...	10–20...	Fixed.
Chart reading, dark adaptation necessary.	Red or white flood with operator choice.	0.1–1.0 (on white portions of chart).	Continuous throughout range.
Chart reading, dark adaptation not necessary.	White flood...	5–20...	Fixed or continuous.

Table C-5. Recommended numeral and letter heights

	Height (in.)*	
Nature of markings	Low luminance†	High luminance‡
Critical markings, position variable (numerals on counters and settable or moving scales)......	0.20 – 0.30	0.12 – 0.20
Critical markings, position fixed (numerals on fixed scales, control and switch markings, emergency instructions).........................	0.15 – 0.30	0.10 – 0.20
Noncritical markings (identification labels, routine instructions, any markings required only for familiarization........................	0.05 – 0.20	0.05 – 0.20

*For 28-in viewing distance. For other viewing distances, increase or decrease values proportionately.
†Between 0.03 and 1.0 ft.-L.
‡Above 1.0 ft.-L.

Table C-6 Relative evaluation of basic symbolic indicator types

For—	Counter is—	Moving pointer is—	Moving scale is—
Quantitative reading.	Good (requires minimum reading time with minimum reading error).	Fair...	Fair.
Qualitative and check reading.	Poor (position changes not easily detected).	Good (location of pointer and change in position is easily detected).	Poor (difficult to judge direction and magnitude of pointer deviation).
Setting...	Good (most accurate method of monitoring numerical settings, but relation between pointer motion and motion of setting knob is less direct).	Good (has simple and direct relation between pointer motion and motion of setting knob, and pointer-position change aids monitoring).	Fair (has somewhat ambiguous relation between pointer motion and motion of setting knob).
Tracking...	Poor (not readily monitored, and has ambiguous relationship to manual-control motion).	Good (pointer position is readily monitored and controlled, provides simple relationship to manual-control motion, and provides some information about rate).	Fair (not readily monitoried and has somewhat ambiguous relationship to manual-control motion).
Orientation...	Poor...	Good (generally moving pointer should represent vehicle, or moving component of system).	Good (generally moving scale should represent outside world, or other stable frame of reference).
General...	Fair (most economical in use of space and illuminated area, scale length limited only by number of counter drums, but is difficult to illuminate properly).	Good (but requires greatest exposed and illuminated area on panel, and scale length is limited).	Fair (offers saving in panel space because only small section of scale need be exposed and illuminated, and long scale is possible).

Ergonomic Data

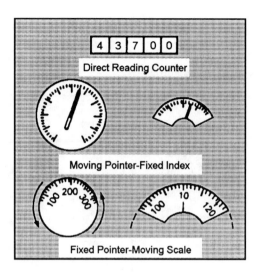

Figure C-1. Basic types of symbolic indicators.

Table C.7. General recommendations for group viewing for slides and motion pictures

Factors	Optimun	Preferred limits	Acceptable limits
Ratio of $\frac{\text{viewing distance}}{\text{image width}}$	4	3–6	2–8
Angle off center line—degrees	0	20	30
*Image luminance—ft.-Lamberts (no film in projector).	10	8–14	5–20
Luminance variation across screen—ration of maximum to minimum luminance.	1	1.5	3.0
Luminance varaiation as a function of seat position—ratio of maximum to minimum luminance.	1	2	4.0
Ratio of $\frac{\text{ambient light}}{\text{highest part of image}}$	0	0.002–0.01	0.2†

*For still projections higher values may be used.
†For line drawings, tables, not involving gray scale or color.

Control Recommendations

Figures C-2 and C-3 involve controls.

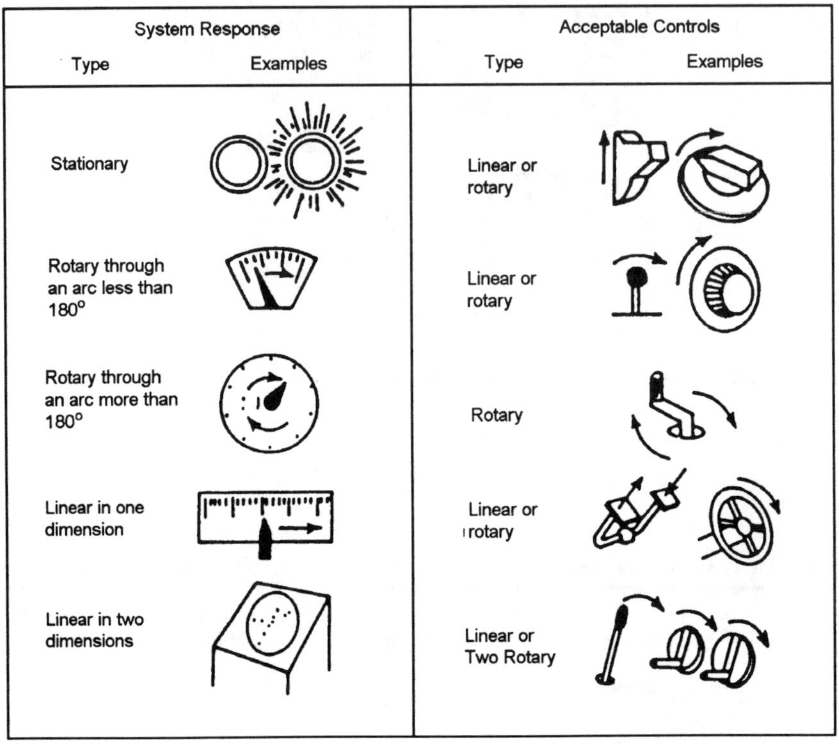

System Response		Acceptable Controls	
Type	Examples	Type	Examples
Stationary		Linear or rotary	
Rotary through an arc less than 180°		Linear or rotary	
Rotary through an arc more than 180°		Rotary	
Linear in one dimension		Linear or rotary	
Linear in two dimensions		Linear or Two Rotary	

Figure C-2. Examples of acceptable controls for various types of system response.

Ergonomic Data

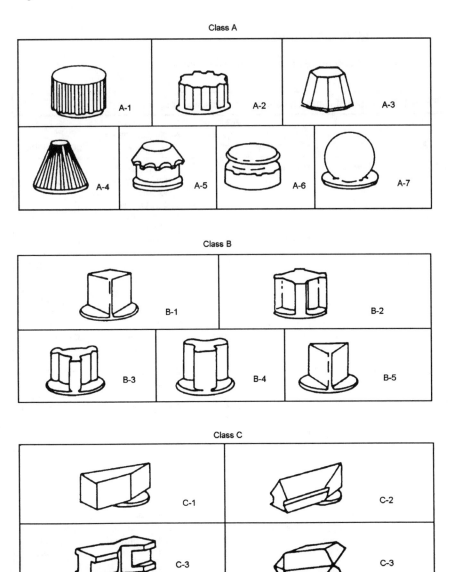

Figure C-3. Examples of three classes of knobs: (A) those for twirling or spinning; (B) those to be used where less than a full turn is required, and position is not so important; and (C) those where less than a full turn is required, and position is important.

Auditory Recommendations

Tables C-8 and C-9 involve auditory considerations.

Table C-8. Types of alarms, their characteristics, and special features

Alarm	Intensity	Frequency	Attention-getting ability	Noise-penetration ability	Special features
Diaphone	Very high...	Very low...	Good...	Poor in low-frequency noise. Good in high-frequency noise.	
Horn...	High...	Low to high.	Good...	Good...	Can be designed to beam sound directionally. Can be rotated to get wide coverage.
Whistle...	High...	Low to high.	Good if intermittent.	Good if frequency is properly chosen.	Can be made directional by reflectors.
Siren...	High...	Low to high.	Very good if pitch rises and falls.	Very good with rising and failing frequency.	Can be coupled to horn for directional transmission.
Bell...	Medium...	Medium to high.	Good...	Good in low-frequency noise.	Can be provided with manual shutoff to insure alarm until action is taken.
Buzzer...	Low to medium.	Low to medium.	Good...	Fair if spectrum is suited to background noise.	Can be provided with manual shutoff to insure alarm until action is taken.
Chimes and gong.	Low to medium.	Low to medium.	Fair...	Fair if spectrum is suited to background noise.	
Oscillator...	Low to high.	Medium to high.	Good if intermittent.	Good if frequency is properly chosen.	Can be presented over intercom system.

Ergonomic Data

Table C-9. Summary of design recommendations for auditory alarm and warning devices

	Conditions		Design recommendations
1.	If distance to listener is great—	1.	Use high intensities and avoid high frequencies.
2.	If sound must bend around obstacles and pass through partitions—	2.	Use low frequencies.
3.	If background noise is present—	3.	Select alarm frequency in region where noise masking is minimal.
4.	To demand attention—	4.	Modulate signal to give intermittent "beeps" or modulate frequency to make pitch rise and fall at rate of about 1–3 cps.
5.	To acknowledge warning—	5.	Provide signal with manual shutoff so that it sounds continuously until action is taken.

Appendix D

Behavioral Research

There is a wealth of research on behavioral factors associated with occupational safety. The problem usually is that it is difficult to find and difficult to understand. A third problem is that what research there is is often done in one industry, and there is uncertainty on whether we can generalize it beyond that industry.

This appendix is an example, a research study performed for the U.S. Bureau of Mines in 1989, by Robert H. Peters. We have included it here because it provides rich information on factors related to safety in the mining industry, which we hope will provide insight to all industry's behavioral safety factors.

REVIEW OF RECENT RESEARCH ON ORGANIZATIONAL AND BEHAVIORAL FACTORS ASSOCIATED WITH MINE SAFETY

By Robert H. Peters[1]

ABSTRACT

This report presents a literature review conducted by the U.S. Bureau of Mines. The review summarizes what has been learned from recent empirical studies of the relationship between mine safety and characteristics of miners, mine management, and mining companies. The review is based on literature published in the United States during the period from 1976 to 1988. In addition to summarizing the empirical findings, this report discusses the implications and limitations of these studies, and gives recommendations for improving mine safety.

---

[1] Research psychologist, Pittsburgh Research Center, U.S. Bureau of Mines, Pittsburgh, PA.

INTRODUCTION

Research on organizational and behavioral factors in the mining industry represents a promising approach to the improvement of miners' safety. Such research projects are usually very expensive and often require several years to perform. Because they are so costly, it is important that researchers begin their studies by reviewing any work that has already been performed. Learning what others have done improves their understanding of the problem and their ability to design an effective study.

The purpose of this U.S. Bureau of Mines report is to provide researchers and mining industry officials with a review of the recent United States research literature concerning the relationship between mine safety and characteristics of miners, mine management, and mining companies. Using computer literature search services, it was found that 17 empirical studies on this topic had been published in the United States during the years 1976 through 1988. For a review of the foreign literature on this topic, see Peters (*23*).[2]

The body of this report is organized into three major sections. The first section reviews the research findings concerning 16 company-level variables; three variables concerning the amount of control employees have over what happens on the job, three variables related to management-labor relations, four variables related to supervisor-employee interactions, and six variables pertaining to individual miners. The second section presents an overview of the findings, and makes some statements concerning the use of caution in interpreting the findings given the nature of the experimental designs and research methods that were used in most of these studies. The third section presents recommendations about improving mine safety that were given by the authors of the 17 studies reviewed in this report. This report also contains an appendix, which gives the following information about each of the 17 studies: the references for publications concerning the study, the study's objective, the methods used to perform the study (including a brief description of experimental interventions and data analysis procedures), the results, and (where applicable) major conclusions.

REVIEW OF RESEARCH FINDINGS

Table D-1 lists the studies included in this review and indicates the following about each: the experimental design, the type of ore mined, whether surface or underground, the number of mines or companies from

---

[2] Italic numbers in parentheses refer to items in the list of references.

Table D-1. Description of empirical studies of organizational and behavioral factors affecting mine safety (1976–1988)

Investigator and reference	Design	Mine type	Sample size	Safety index
Althouse (1)..	Contrast.....	Underground coal...	29 mines....	Not available.
Bell (3)......	Intervention..	Underground silver..	1 mine......	Lost time.
Bennett (4)...	Correlational.	Underground coal...	All U.S. coal..	Lost time, non-lost time.[1]
DeMichiei (9)..	Contrast.....	.. do............	40 mines....	Not available.
Edwards (10)..	Intervention..	Surface coal........	1 mine......	All.
Fiedler (12)...	.. do......	Underground trona...	.. do......	All, violations.
Gaertner (14)..	Correlational.	Underground coal...	10 companies	Do.
Goodman (16).	Intervention...	.. do............	1 mine......	All, violations.[2]
Goodman (17).	.. do......	.. do............	4 mines.....	Lost time.
Goodman (18).	Correlational.	.. do............	5 mines.....	All.
NAS (21):				
1st data set..	.. do......	.. do............	All U.S. coal..	Multiple.[3]
2d data set...	.. do......	.. do............	19 companies	Do.[3]
3d data set...	Contrast.....	.. do............	12 mines....	Lost time.
Peters (24)...	Intervention..	Surface gold	1 mine......	All, hazards.
Do.........	.. do......	Surface coal	2 mines.....	Do.
Do.........	.. do......	Underground coal...	.. do......	Do.
Pfeifer (25)....	Contrast.....	.. do............	28 mines....	Not available.
Rhoton (26)...	Intervention..	.. do............	1 mine......	Hazards.
Sanders (27)..	Cross-lag panel.....	.. do............	22 mines....	Lost time.
Uslan (30)....	Intervention..	Underground salt...	4 mines.....	Selected.[4]
Wagner (31)...	Correlational.	Surface iron........	10 mines....	Days lost.

NAS National Academy of Sciences.
[1] Dependent variable was a dichotomous indicator of whether the injury was considered a lost-time or non-lost time incident.
[2] Five measures from Government safety inspectors were used: total MSHA violations per section, total State violations per section, number of times section was closed down, quarterly global ratings of section safety by Federal inspectors, and quarterly global ratings of section safety by State inspectors.
[3] Rates for several categories of injuries and accidents were computed, including fatalities, lost-time incidents, non-lost-time incidents, and serious incidents.
[4] Total number of eye, head, hand, and back injuries per month.

which data were collected, and the manner in which mine safety was measured.

The experimental design. The studies that look for naturally occurring relationships between variables and measures of mine safety by collecting data at only one point in time are labeled correlational. One study used a cross-lag panel design (*27*). Data were collected at two points in time, and the effect of changes in the correlates on changes in safety measures were examined. This is a relatively good design because it allows one to make better inferences about the direction of causality. In the correlational and cross-lag panel studies, the effects of planned changes on mine safety were not studied. Other studies, labeled interventions, involve observing the effects of some planned change (e.g., a new training program) on indexes of mine safety. Measures of mine safety are taken before and after the intervention is introduced, and inferences are made about the role of the intervention in producing observed changes in safety. Unfortunately, given the frequent lack of adequate control sites, and the complexity of the interventions, it is often quite difficult to determine precisely what was responsible for the observed improvements in safety. In instances where control sites are used, this design allows one to make relatively strong inferences about the existence of causal relationships between variables.

A technique often used in performing research on company-level determinants of employee safety is to look for factors that differentiate between companies that have established records of good versus poor safety performance (7). This type of study is labeled the contrast design. Typically, data are collected on a very large number of factors, and statistical tests are performed to see which factors differentiate between the two groups of companies at a statistically significant level. There are some important limitations to this type of study: (1) When considering a large number of factors, one should expect to observe a certain number of statistically significant relationships between variables that, in reality, are not related to one another; (2) one cannot specify the direction of causality, or the extent to which two variables are influencing each other in a reciprocal manner; (3) one cannot rule out the possibility that the observed relationship might be because of the influence of some unmeasured variable that is highly correlated with the ones measured. In other words, it may be the case that the measured variable is not the true cause of differences in safety performance; rather, the true cause influences both safety performance and the variable found to be correlated with safety performance.

Safety indexes. The most commonly used measure of mine safety is the rate of lost-time injuries reported to the Mine Safety and Health Administration (MSHA) per 200,000 employee hours of exposure. Some studies used the rate of total injuries reported to MSHA per 200,000 employee hours of exposure. This measure is designated by "all." Some critics argue that the lost-time injury rate is a more valid index for assessing differences in safety performance than the "all" injury rate because there could be significant variations in the extent to which mine operators actually report non-lost-time injuries to MSHA. Some studies have defined the measure of safety more narrowly, and have used the rate of injuries to selected body parts, etc. Another measure of safety commonly used is the severity of the injury. Severity is operationally defined as the number of workdays missed following the injury.[3] Another measure of safety is the number of violations issued

---

[3] Although none of the studies in this review used it, another potentially useful measure of accident severity would be an index that normalizes the number of workdays lost by the number of hours worked, e.g., number of lost workdays per 200,000 employee hours.

by mine inspectors. Finally, some studies have defined safety in terms of the number of hazardous conditions or instances of unsafe work practices observed during periodic inspections of the worksite.

Except for the studies by Edwards (10), Goodman (15–18), Peters (24), and Uslan (30), all these studies used data provided by MSHA to compute rates of accidents, injuries, and violations, and to define injury severity. Severity is usually defined as the number of days away from work following the injury. With the exception of the studies by Bennett (4), Edwards (10), and Rhoton (26), all the studies were funded by agencies of the U.S. Government, primarily the Bureau.

Table D-2 lists the organizational and behavioral variables that have been found related to some measure of mine safety at the probability (p) < 0.05 level of statistical significance. Positive relationships to mine safety should be interpreted as: increases in variable X (or the presence or use of X) are associated with a better safety record. Negative relationships to mine safety should be interpreted as: increases in variable X (or the presence or use of X) are associated with a poorer safety record. The variables are organized into five categories: organizational, employee control, management-labor relations, supervisor-employee interaction, and individual miner. The empirical evidence concerning each of the variables listed in Table D-2 is discussed in the following pages.

ORGANIZATIONAL FACTORS

This category of variables includes factors other than management-labor relations that characterize or that affect the entire mining operation or company, such as the size of the operation, management's emphasis on achieving various objectives, use of safety policies and practices, company-wide training programs, etc. These variables describe organizational level characteristics (as opposed to characteristics of individuals or workgroups within the organization) that have been examined in studies of mine safety performance.

Training for Miners

DeMichiei (9) made the following observations of significant differences in the mean responses of mine managers and safety officials from high- versus low-accident rate mines: (a) new miners in high-accident rate mines were less informed on how to do their jobs than new miners in low-accident rate mines, (b) supervisors at high-accident rate mines did not provide the same degree of instruction and guidance to miners as did supervisors in low-accident rate mines.

Pfeifer (25) made the following observations of significant differences in the mean responses of miners from high- versus low-accident rate mines: (a) there is a consistent tendency for training on various topics to be rated as better

Table D-2. Summary of organizational and behavioral variables associated with mine safety

Variable	Investigator and reference	Association direction
Organization:		
Training for miners	DeMichiei (9)	Positive.
Do	Pfeifer (25)	Do.
Training for managers and supervisors.	Fiedler (12)	Do.[1]
Do	Peters (24)	Do.[1]
Do	Uslan (30)	Do.[1]
Management planning	DeMichiei (9)	Do.
Management commitment to safety	.. do	Do.
Do	Pfeifer (25)	Do.
Production competition between crews	.. do	Negative.
Production pressure	.. do	Do.
Do	Sanders (27)	Do.
Equipment availability	DeMichiei (9)	Positive.
Safety & health equipment maintenance	Pfeifer (25)	Do.
Safety-production incentive program	Gaertner (14)	Do.
Safety disciplinary actions	.. do	Do.
Size of mine	NAS (21)	Do.
Feedback and praise	Rhoton (26)	Do.[1]
Night shiftwork	Wagner (31)	Negative.[2]
Employee control:		
Worker participation in problem solving	Edwards (10)	Positive.[1]
Do	Bell (3)	Do.[1]
Do	DeMichiei (9)	Do.
Worker autonomy	Goodman (16)	Do.[1]
Do	Sanders (27)	Do.
Decentralized decisionmaking	DeMichiei (9)	Do.
Management-labor relations:		
Overall labor relations climate	Gaertner (14)	Do.
Management concern for labor	DeMichiei (9)	Do.
Do	Pfeifer (25)	Do.
Labor support for safety discipline	DeMichiei (9)	Do.
Supervisor-employee interaction:		
Reporting hazards to supervisor	Pfeifer (25)	Do.
Employee development	Sanders (27)	Do.
Praise for working safe	Pfeifer (25)	Do.
Do	Uslan (30)	Do.
Communications to miners	DeMichiei (9)	Do.
Individual miner:		
Absenteeism	Goodman (18)	Negative.
Do	DeMichiei (9)	Do.
Do	Pfeifer (25)	Do.
Coworker relations	.. do	Positive.
Role ambiguity	.. do	Negative.
Do	Althouse (1)	Positive.
Role overload	Pfeifer (25)	Negative.
Role conflict	.. do	Do.
Do	Sanders (27)	Do.
Age	NAS (21)	Positive.

[1] Variable was only one of several changes introduced more or less simultaneously as part of a complex organizational intervention. It is not clear how much, if any, of the observed effects on safety should be attributed to this variable.
[2] Accidents experienced on night shift are more severe than those experienced on day or afternoon shifts.

better by underground miners at low-accident mines, (b) good training in how the electrical power system works, dealing with hazards (such as gases, coal dust, and noise), and how to use tools and equipment is more prevalent in mines with lower accident rates, and (c) lack of training in the proper use of safety and health equipment was more frequently cited as being an important reason for not using the equipment by miners at high-accident mines.

Training for Managers and Supervisors

A major focus of three intervention studies that produced significant improvements in safety was the provision of training for managers and supervisors. Fiedler (*11*) used a training intervention called the Leader Match program. This program teaches individuals to diagnose their own leadership styles, as well as to diagnose the leadership situations. The leaders are given detailed instruction on various methods for modifying the situations to match their particular management approaches and personalities. The instruction is provided by a trainer, who uses a detailed manual, aided by videotaped illustrations, slides, and/or transparencies. Videotaped vignettes in actual and staged settings were used to teach supervisors how to deal with specific problems with employees including: reinforcing safe behavior, correcting an employee, overcoming resistance to change, handling an irate employee, and creating a cooperative workteam.

The training intervention reported by Peters (*24*) entailed (a) conducting 8 h of training for subordinate managers and supervisors in loss-control fundamentals; (b) providing 8 h of training to supervisors and workers selected by the top manager in the use of an accident investigation methodology developed by the Bureau; and (c) furnishing available materials for training mobile equipment operators and assisting in task training course development.

The training intervention reported by Uslan (*30*) was called positive motivational safety training (POMOST). This training provides supervisors with "an understanding about behavior and a process by which they can improve performance; increase safe work behaviors, help the employee feel better about himself, the company, and his job; and generally establish a work environment which is positively supportive." The principal reinforcer used in POMOST is praise, i.e., positive verbal feedback. POMOST instructs trainees how and when to praise their employees. The program was taught in 22 to 40 h, depending on need. The program objectives were as follows:

How to recognize unsafe behaviors.
How to develop behavior baselines.
How to determine what behaviors to change.
How to communicate behavior change to employees.

How to shape behaviors.
How to maintain a safe work behavior program.

This training discourages the use of punishment. Rather, emphasis is placed on the first line supervisor focusing the employees' attention on the appropriateness of their behavior. The existence of inappropriate behavior is not ignored, but perceived as a training problem.

Approximately 100 managers and supervisors from four plants were trained in the use of positive reinforcement for occurrences of safe behavior. Each supervisor was also provided manuals and other supportive materials. Additionally, following training, all supervisors were provided some coaching experience to keep the knowledge and capabilities gained during training from being lost.

Management Planning

DeMichiei's (9) questionnaire results suggest that in comparison with low-accident rate mines, there was a greater tendency for management to put off making important decisions and more worktime was lost through poor scheduling and planning at high-accident mines.

Management Commitment to Safety

Pfeifer's (25) questionnaire data suggest that in comparison with miners in high-accident mines, miners in low-accident mines felt that keeping good safety records and upholding the company safety record were significantly more important to their companies.

DeMichiei's (9) questionnaire results suggest that management at high-accident rate mines provided less support for decisions made by section supervisors concerning safety than did management at low-accident rate mines.

DeMichiei (9) makes the following observations regarding information provided to him through interviews with managers and safety officials: (a) in five high-accident rate mines, the mine superintendents had no direct involvement in the mine's safety and health program; (b) responsibility for implementing the program was mainly the safety department's; and (c) safety department personnel at these five high-rate mines identified the lack of upper management involvement in safety matters as a serious impediment to improving safety and health conditions at the mine.

Production Competition Between Crews

Pfeifer's (25) questionnaire results suggest that in comparison with miners in high-accident mines, miners in low-accident mines felt that having competition among workers was significantly less important to their companies.

Production Pressure

Pfeifer's (25) study suggests that in comparison with supervisors in high-accident mines, supervisors in low-accident mines are significantly less inclined to push hard for production or to cut corners on safety.

Sanders' (27) findings, based on a cross-lagged panel design, suggest that a causal relationship exists between increased levels of production pressure and increases in the rate of lost-time injuries. Sanders states, "Production pressure appears to lead to an increase in disabling injuries which in turn results in a decrease in production pressure."

Equipment Availability

DeMichiei's (9) questionnaire study suggests that miners in low-accident rate mines are better provided with supplies, equipment, and the tools necessary for job accomplishment than are miners at high-accident rate mines.

Safety and Health Equipment Maintenance

Pfeifer's (25) questionnaire results indicate that miners in low-accident rate mines cite poor maintenance of health and safety equipment as being an important reason for not using it significantly more often than miners in high-accident rate mines.

Safety-Production Incentive Program

Although the specific terms of combined safety-production incentive plans vary, they all have the same basic structure. These plans pay out some form of bonus for production above a specified target, provided the number of accidents does not exceed some specified maximum number. If more than the maximum number of accidents occur, then bonuses are not paid for exceeding the production target, or the amount of the bonus is reduced according to some predetermined formula.

Gaertner (14) compared coal companies using a combined safety-production incentive program with companies not using such a program. He found that the average rate of lost-time injuries was lower for the companies that were using the combined safety-production incentive program. Likewise, Page (22) performed a case study of the characteristics common to high producing underground coal mines and found that 18 out of 25 of them had some form of combined safety-production incentive program.

Gaertner (14) also compared coal companies using an incentive program based on safety performance alone with companies that had no type of incentive program, and found that the average rate of injuries was essentially the same for the two groups of companies. Likewise, Goodman (*17*) found that the rate of accidents was NOT significantly affected by the introduction of a safety incentive program at any of the four different coal mines included in the study.

The reasons for the observed difference in safety performance between companies using combined safety-production incentive programs versus those using an incentive program based on safety performance alone are not entirely clear. Gaertner offers the following speculation concerning his results:

> Safety incentives not tied to production tended to be relatively inexpensive and symbolic (T-shirts, caps, decals) and generally did not indicate a serious commitment by management to safety in the eyes of the foremen and hourly employees. Rather they were usually intended as "reminders to the men, to keep thinking about safety," as one manager put it. Combined production-safety incentives seem more effective, and generally more costly. In one company, the cost of the production-safety incentives was 25 cents per ton; in another it was 81 cents per ton. As one foreman put it, the incentive program in place in his company "made management put their money where their mouth was."

Safety Disciplinary Actions

Some coal companies have (and use) disciplinary actions applied to safety violations by miners and supervisors. Gaertner (*14*) compared companies using such a policy with companies not using such a policy, and found that the rate of injuries was lower at companies that used a safety discipline policy.

In Braithwaite's (*5*) case study of the five major coal companies with the best safety records during the early 1980's, he notes that there was considerable variation in their use of punitive actions for violation of safety policies. Two of the five companies used punitive measures much more often than the other three.

Size of Mine

The National Academy of Sciences (*21*) analyzed MSHA's data concerning all lost time injuries that occurred in U.S. underground coal mines during the 3-yr period 1978 through 1980; such injuries numbered nearly 40,000, including fatalities. One of the variables, which emerged as highly correlated with rate of fatalities, was mine size. It was found that the fatality rate for mines with 50 or fewer employees (0.14) was about three times that of mines with over 250 employees (0.05), and almost twice that of mines with 51–250 employees. Small mines (< 50) accounted for 15 pct of total employee-hours, but 40 pct of all fatalities.

Feedback and Praise

Rhoton (26) was successful in virtually eliminating MSHA ventilation violation notices at a coal mine that, prior to the intervention, had been experiencing an average of 2.6 such violations per month. The intervention was conducted by the company's safety inspector, and consisted of: praising face crews and their supervisors for being in compliance with the target behaviors; posting graphic feedback charts of the ventilation violation notices in the mine office; delivering verbal feedback at biweekly safety meetings with the supervisors; and when a section was found to be noncompliant, the company safety inspector would stop coal production until the hazardous condition was corrected. Each section was observed once per week on a random basis.

Night Shift Work

Research by Andlauer (2) suggests that workers performing routine tasks during periods of low alertness and arousal will tend to introduce few errors into these tasks. However, the inhibited state of psychological arousal coupled with a narrow conscious focus on the routine task at hand may not allow the employee to respond properly to emergencies, thereby promoting the causation of relatively severe accidents.

Wagner (31) examined this issue by comparing the average severity of night shift accidents that involved the operation of heavy mining equipment versus accidents involving miners performing other types of activities—ones that presumably require less vigilance to perform the job safely. The data for this study consisted of accidents reported to MSHA from 10 surface iron ore operations over a 10-yr period. These mines all worked an identical shift rotation schedule for that full 10-yr period.

Accidents were broken down according to the shift (day, afternoon, night) and according to whether the accident occurred during equipment operation (selected accidents) or during the course of performing some other type of activity (nonselected accidents). The average number of days lost per accident was used as the measure of accident severity.

Wagner (31) found that the average length of time to recover from mobile equipment accidents was significantly greater for accidents experienced during the night shift than for accidents experienced during the day or afternoon shifts. Among nonselected accidents, night shift accidents are also more severe, but by a much smaller margin compared with the other two shifts. These data appear to confirm findings by Andlauer (2) and others regarding automatic behavior and employees' inability to respond adequately in emergency situations during times such as the night shift when vigilance is difficult to maintain.

EMPLOYEE CONTROL

Three variables have been studied that concern the extent to which employees have the opportunity to influence what happens to them on the job: (1) worker participation in problem solving, (2) worker autonomy, and (3) decentralized decision making. Although the meaning of worker participation in problem solving is rather self-evident, the latter two variables may not be as easily understood. Definitions of autonomy include: the amount of control employees have over what happens on the job; the extent to which employees see themselves as free to do what they want in accomplishing their work; perceived self-determination (*13*). One way of increasing perceived autonomy is by increasing employees' opportunity to influence decisions that affect them.

Although the notion of decentralization can be applied to hourly employees, it is more typically used to characterize decision making at higher levels in the organization. Webber (*32*) refers to decentralization as the extent to which higher executives delegate authority to make decisions to subordinate managers.

Worker Participation in Problem Solving

Bell (*3*) reports that injury rates at an underground silver mine were significantly lower following the institution of an organization development effort. A primary feature of this intervention was the formation of problem solving teams. A series of meetings took place in which high-priority issues facing the team were systematically examined and resolved. These meetings usually were conducted with the aid of a consultant who acted as a facilitator. Problems were defined and clarified, alternative solutions were evaluated, preferred solutions were implemented, and the effects of actions were monitored for desired results. Team-building meetings began at the top of the organization, involving the president and those reporting to the president. During the last phase of the project, team-building meetings were held with shift bosses and their workcrews.

Similarly, Edwards (*10*) reports that the introduction of quality circles at a large surface coal mining operation brought about a significant reduction in the accident rate.

DeMichiei's (*9*) questionnaire results (based on the responses of mine managers and safety officials) suggested that good ideas fostered by miners get more serious consideration from management at low-accident rate mines than at high-accident rate mines.

DeMichiei (9) makes the following three observations based on interviews with miners:

(1) At low-accident rate mines, miners indicated that individuals in key management positions were receptive and responsive to miner requests and frequently solicited input from the miners concerning mine policies and procedures.

(2) At several low-accident rate mines, miners indicated that management earnestly encouraged interaction between themselves and labor representatives. Management at most of the low-accident rate mines recognized the importance of labor safety and health committees and actively sought out their participation in resolving safety and health problems.

(3) Miners at several high-accident rate mines believed management to be one-sided since they had little input into the decision making process.

Worker Autonomy

The results of the autonomous workgroups experiment at Rushton Mining Co., Osceola Mills, PA, suggests that increasing miners' autonomy may have been responsible for the improvements that were observed in mine safety. Although this intervention entailed making a number of changes to the organization, one of the most notable changes was the creation of self-managing coal mine face crews. These crews were given the entire responsibility for making all the decisions pertaining to the day-to-day production of coal from their sections. Crew supervisors were no longer involved in making production decisions. Their primary responsibility was to maintain the safety of the crew.

It is actually very difficult to determine what role autonomy played in producing the safety improvements. Goodman (16) notes that any of the following changes could have been responsible for the safety improvements:

(1) The experimental group received more formal training about safety practices and the law.

(2) The experiment introduced a new regard system to motivate the workers toward good safety practices. Intrinsic rewards were increased: The workgroup had been restructured to provide the workers greater opportunities for feelings of responsibility and accomplishment if safety levels improved. Extrinsic rewards were also increased through formal feedback sessions for the workers concerning their performance on such activities as safety.

(3) The supervisors in the experimental section were no longer caught in the

conflict between production and safety. They could expend all their energies on safety. Supervisors in the nonexperimental sections had continually to balance production and safety demands.

Sanders' (27) findings, based on a cross-lagged panel design, suggests that a causal relationship exists between greater decentralization of decision making (down to the miners' level) and a lowering of the rate of lost-time injuries. He states, "mines in which miners are given decision responsibility and autonomy tend to have a lower incidence of injuries than other mines."

Decentralized Decision Making

DeMichiei's (9) questionnaire results (based on the responses of mine managers and safety officials) suggest that section supervisors at high-accident rate mines did not have as much freedom to make decisions concerning health, safety, and production as did section supervisors at low-accident rate mines.

Braithwaite's (5) case study of five coal companies with outstanding safety records suggests that decentralization of decisions regarding safety is a characteristic that is common among the large mining companies with better safety records.

These results concerning a positive association between decentralized decision making and mine safety do not contradict those discussed earlier concerning the value of upper management's interest and commitment to ensuring their miners' safety. As Gaertner (14) points out, upper management can express this commitment through setting safety-oriented policies and showing support for those charged with the responsibility for seeing to it that safety policies are being implemented and adhered to, as well as by frequently reviewing safety performance and generally keeping abreast of what is being done to maintain and improve safety. However, this does not mean that upper management should be running the show. Braithwaite notes that in mining companies with better safety records, upper management tends to delegate the responsibility for implementation and adherence to safety policies to line management. The fact that upper management delegates this function does not mean that they are not keeping a close watch on safety performance or that they are disinterested in employee safety.

MANAGEMENT-LABOR RELATIONS

This category includes the variables: overall labor relations climate, management concern for labor, and labor support for safety disciplinary actions. A considerable amount of evidence is accumulating to suggest that there is a significant positive association between the favorableness of the management-labor relationship and mine safety. Part of this association reflects the tendency for good labor relations to result in a safer environment and better employee compliance with safety rules. However, part of this association may reflect the tendency for changes in safety to produce changes in management-labor relations. For instance, a deterioration in safety could tend to cause a deterioration in management-labor relations. No doubt there is a reciprocal relationship between these two factors. Given the type of research evidence available, it is not possible to specify the predominant direction of causality. Nevertheless, the fact that a consistent positive association between these two factors has been identified is worth noting and has important implications for the achievement of a safer mining industry.

Overall Labor Relations Climate

Gaertner's (14) analysis of factors correlated to coal companies' safety performance indicates that the injury rate in companies with a negative labor relations climate is almost double that of the rate in companies with a positive climate. He also compared the two groups of companies in terms of the average annual rate of MSHA citations per mine for significant and substantial safety violations. Gaertner found that the rate for companies with positive labor relations was only about a third as high as the rate for companies in which the labor relations climate was relatively negative.

As part of the National Academy of Sciences (21) study, interviews were conducted with miners and mine managers at twelve large underground coal mines (150 workers or more). These mines were selected partially because they represented the extremes of safety performance, i.e., seven had quite low injury rates and five had quite high injury rates. The researchers note the following concerning the labor relations climate at these two groups of mines: "at all seven mines with low injury rates there appeared to be a cooperative attitude between management and labor; an adversarial attitude was observed in three of the five mines with high injury rates."

Both Gaertner (14) and the National Academy of Sciences (21) researchers point out that managers at mines with favorable labor relations usually had an open door policy, i.e., they were more accessible to hourly employees.

Management Concern for Labor

DeMichiei's (9) questionnaire results suggest that management at high-accident rate mines is less interested in the welfare of workers both on and off the job than management at low-accident rate mines. These results are based on the perceptions of salaried employees (includes mine superintendents, section supervisors, and mine safety department personnel).

Pfeifer's (25) questionnaire results suggest that supervisors in low-accident rate mines "more often show real concern for workers' welfare." These results are based on the perceptions of underground hourly employees.

Labor Support for Safety Discipline

DeMichiei's (9) questionnaire results suggest that section supervisors in low-accident rate mines received more support from safety committeepersons when reprimanding miners for unsafe acts. Based on interview data with miners he states,

> A significant finding at 11 high-accident rate mines was statements made by safety and health committees concerning their reluctance to support management's decision to reprimand miners for unsafe acts, even though such actions were justified. They stated that if such support was given to management, the rank and file miners usually reacted in a hostile manner. Management believed that such reaction by labor representatives prevented a good faith effort on the part of both parties to promote health and safety at the mines.

This same issue was examined in the part of the National Academy of Sciences (21) study that compared underground coal mines with either very low- or very high-accident rates. The researchers note that at all of the seven low-injury rate mines the union generally supported the company's enforcement of safety rules.

SUPERVISOR-EMPLOYEE INTERACTION

The miners' immediate supervisor has a great deal of impact on the extent to which safe working conditions exist and the extent to which miners comply with safety regulations. Variables dealing with the supervisor-employee relationship that have been addressed by mining industry safety researchers include: reporting hazards to supervisor, employee development, praise for working safely, and communications to miners. No doubt many of the variables listed in the other categories are also heavily influenced by the immediate supervisors' behavior, e.g., management-labor relations and role conflict.

Reporting Hazards to Supervisor

Pfeifer's (25) questionnaire results suggest that miners in low-accident rate mines reported small accidents and safety and health hazards to the supervisor more often than miners in high-accident rate mines. Similarly, DeMichiei's (9) questionnaire results suggest that miners contacted section supervisors regarding unsafe conditions more often at low-accident rate mines than at high-accident rate mines.

Employee Development

Sanders' (27) questionnaire results suggest that supervisors at mines with low-accident rates did more to instruct, guide, coach, and develop their employees' talents and abilities than supervisors at mines with high-accident rates.

Praise for Working Safely

Although the intervention involved several other changes, one of the primary things Uslan (30) taught the supervisors of salt mine employees to do was to systematically praise their employees for working safely. A significant reduction in the rate of eye, head, hand, and back injuries took place following the intervention.

Similarly, Pfeifer's (25) questionnaire results suggest that in comparison with high-accident rate mines, supervisors in low-accident rate mines were more likely to give miners a pat on the back when they followed safety procedures.

Communications to Miners

DeMichiei's (9) questionnaire results suggest that there is more conflict or misunderstanding over directions and job assignments given at high-accident rate mines.

INDIVIDUAL MINER

Each of the variables in this category characterize a single person or a single person's workrole. Variables in this category that have been addressed by safety researchers in the mining industry include: absenteeism, coworker relations, role ambiguity, role overload, role conflict, and age. No doubt the three variables concerning the individual's role are also heavily influenced by the immediate supervisor and by some of the variables listed under the organizational category, e.g., training, feedback, and management planning.

Absenteeism

Goodman (18) examined the effects or consequences of absenteeism on accidents. His central premise is that lack of familiarity leads to more dangerous conditions that, in the absence of compensatory changes in the level of care taken by a miner, would contribute to higher rates of accidents. Unfamiliarity can affect three types of people in coal mining: 1) miners who have just returned to the mine after being absent, 2) miners assigned as a replacement for a miner who is absent, and 3) adjacent miners. In a typical crew configuration, most mining activities require coordination among pairs of individuals who work closely together: the miner operator and the helper, the roof bolter and the helper, and the two shuttle car operators. Considering such dyads, the worker adjacent to the replacement may also be placed in a more dangerous situation because of lack of familiarity with the replacement's mining practices, and resulting difficulties in coordinating activities in an inherently dangerous environment. Hence, this adjacent worker or partner of a replacement is expected to have a higher probability of having an accident.

A data set was created that kept daily records on absences, replacement policies, and accidents. Information from five mines on these variables was obtained during approximately 70,000 miner days worked. This corresponds to roughly 60 production crews observed for an entire year. Using multiple regression analyses, accident rates were computed and comparisons made between each of the following 11 dyad-familiarity categories:

A   Two regulars, neither absent previous day.
B   Two regulars, one absent previous day.
C   Two regulars, both absent previous day.
D   Two replacements, neither absent previous day.
E   Two replacements, one absent previous day.
F   Two replacements, both absent previous day.
G   One regular, one replacement, neither absent previous day.
H   Regular absent previous day, replacement present previous day.
I   Regular present previous day, replacement absent previous day.
J   One regular, one replacement, both absent previous day.
K   One miner working without partner.

Goodman (*18*) found that in comparison with category F, the accident rates for categories A, C, and D were each significantly lower. Most of the other expected differences between pairs of accident rates were in the expected direction, but were relatively small in magnitude. Goodman concludes that prior day absences have the effect of increasing accidents, and that regular miners have lower accident rates than replacement miners.

DeMichiei's (*9*) questionnaire results suggest that absenteeism is much more of a problem at high-accident rate mines than at low-accident rate mines.

Absenteeism was one of the issues examined in the part of the National Academy of Sciences (*21*) study that compared underground coal mines with either very low- or very high-accident rates. The researchers note that the five mines with high-injury rates tended to have considerably higher absenteeism rates.

The responses miners and supervisors gave in Pfeifer's (25) questionnaire study of the differences between high- and low-accident rate mines suggest that miners in low-accident mines had to take care of absentees' jobs significantly less often than miners in high-accident mines.

Coworker Relations

Pfeifer's (25) questionnaire results suggest that in comparison with the high-accident mines, miners in low-accident mines are more apt to believe that miners get along with each other and that they can depend on each other.

On the basis of interviews with miners, DeMichiei (9) reports that in comparison with low-accident mines, more miners at high-rate mines indicated that they were troubled by some of their coworkers' behavior. These complaints were centered around "freeloaders who often took advantage of disability compensation, individuals who failed to pull their share of the workload and persons who abused absentee policies for personal gain."[4]

Role Ambiguity

Pfeifer's (25) questionnaire results suggest that in comparison with the low-accident mines, miners in high-accident mines spend more time working without clear-cut duties.

Althouse's (1) findings suggest just the opposite. He collected data concerning the extent to which coal miners experience 30 separate dimensions of job stress from the same sample of mines that Pfeifer (25) studied. Very little evidence of any differences in the self-reported levels of stress experienced by miners employed at high- versus low-accident rate mines was found. However, contrary to the hypothesis—that miners at high-accident rate mines would report higher levels of perceived role ambiguity than miners at low-accident rate mines—Althouse (1) found just the opposite. The *low* accident, rather than the high-accident sample reported significantly higher levels of role ambiguity. In other words, miners in low-accident mines were more likely to report that the clarity of responsibility and certainty of

---

[4]DeMichiei did not perform statistical tests on differences between the *interview* responses of various personnel from high- versus low-accident mines. Therefore, such differences are not included in Table D-2. DeMichiei did perform statistical tests on differences between the *questionnaire* responses of various personnel from high- versus low-accident mines, and those differences found significant (p < .05) are listed in Table D-2.

objectives on their jobs was ill-defined. Obviously, more research evidence is needed in order to form any conclusions about the relationship between miners' role ambiguity and safety.

Role Overload

Pfeifer's (25) questionnaire results suggest that in comparison with the high-accident mines, significantly fewer miners at low-accident mines report that they are expected to do too many things in too little time.

Role Conflict

Pfeifer's (25) questionnaire results suggest that in comparison with the high-accident mines, significantly fewer managers in low-accident mines felt that miners had to answer to more than one person.

One of the variables considered in Sanders' (27) cross-lag panel study of mine safety was miners' ratings of the consistency of the orders they were being given. Sanders (27) found that increases in the lost-time injury rate tended to bring about increases in the rated consistency of orders, which then appeared to decrease the lost-time injury rate ($p < .10$).

These findings concerning miners' perceptions of role ambiguity, overload, and conflict may all be indications of insufficient management planning at the high-accident rate mines.

Age

The National Academy of Sciences (21) study looked at the relationship between age and various accident rates for 15 of the largest underground coal producing companies in the United States. These companies provided data to the researchers concerning the age of their work force, and the accident data came from the reports these companies filed with MSHA concerning lost-time injuries that occurred in their mines during the 3-yr period 1978 through 1980. The researchers found that there is no evidence of an age trend with respect to fatality rates or permanent disability injury rates. However, there is a very marked correlation between age and lost-time injury rates. Miners between the ages of 18 and 24 have a lost-time injury rate nearly twice that of miners 25 to 34, who have a rate about 25 pct higher than miners 35 to 44, who in turn have a rate over 40 pct higher than miners who are at least 45 yr of age. Hence, a young miner (18–24) is about three times more likely to be injured than is a miner 45 yr of age or older, about twice as likely to be injured than is a miner 25–44. This relationship is consistent across the 15 companies that provided data on the age of their work force, as well as for each of the years 1978, 1979, and 1980. Furthermore, the strong association between age and lost-time injury rate was apparent for each of the major categories of accident types that cause injuries.

Findings of nonsignificant relationships to mine safety. Many examples of nonsignificant relationships between mine safety and various organizational and behavior variables have been reported. Because these variables are so numerous, they will not all be listed. The interested reader should examine the results of the analyses of responses to questionnaire items reported in the large survey studies conducted by DeMichiei (9), Pfeifer (25), and Sanders (27).

Two variables have been examined in a relatively thorough manner and very little evidence was found to suggest that they are related to differences in mine safety performance: (1) Althouse (1) found that self-reported levels of job stress and psychological strain were NOT significantly different for miners at 14 high- versus 14 low-accident rate coal mines. (2) Goodman (17) found that the rate of accidents was NOT significantly affected by the introduction of a safety incentive program at four different coal mines. Likewise, Gaertner (14) found that there was essentially no difference between the injury rate for companies using versus not using a safety incentive program. It may be that safety incentive programs need to be designed differently than those examined by Goodman (17) and Gaertner (14) in order to have a significant impact on miners' injury rates. Goodman suggests several ways that the plan studied could have been made more effective. It may be well worth having a safety incentive program even if it does not by itself cause statistically significant reductions in the injury rate. Safety incentive programs represent one of several mechanisms that management could use to communicate their concern for employees' welfare and the fact that they want their employees to focus on working safely. For a discussion of mine operators' views on safety incentive programs see Miles (19).

OVERVIEW OF FINDINGS AND THEIR LIMITATIONS

Given the nature of the experimental designs and the data collection methods used to conduct most of the research presented in this review, there are some serious limitations to what can be safely concluded from any one study about how an organizational or behavioral factor influences mine safety. In most of these studies several alternative explanations exist for the observed correlations or for the observed changes in safety performance. Future studies of mine safety should seek to use longitudinal designs, control groups, and multivariate analyses. Two significant limitations to performing the types of research that would yield the most convincing results are (1) the length of time allowed for completing research projects is typically insufficient to permit analyses to be performed that are based on a longitudinal experimental design, and (2) it is difficult to find companies that are willing to participate in the research, especially if the research will require extensive data collection or entails significant changes to well-established practices.

Another limitation of the current set of research findings concerns the extent to which these findings can be generalized to all segments of the mining industry. The bulk of the research has been conducted at large underground coal mines and large coal mining companies. Therefore, it may be inappropriate to assume that these results would also be true of smaller mining operations, surface mining operations, or noncoal mining operations. The best management practices for achieving a good safety record at large mining operations may not be feasible or practical for relatively small mines. Conversely, the best management practices for achieving a good safety record at small mining operations may not be practical for relatively large operations. Future research should be performed to discover why the fatality rate is so much higher at small operations, and to discover what accounts for differences between small mine operations that have, versus those that have not been successful at maintaining a good safety record.

Certain variables are emerging as statistically significant correlates of mine safety in multiple studies involving different samples of miners and different research methods. For instance, it appears that better labor-management relations, greater employee involvement in decision making, and lower absenteeism all exert a positive influence on mine safety. Several aspects of first line supervisors' interactions with miners also appear to be important, as well as management's ability to communicate to the miner that they truly consider the employees' safety and welfare as a top priority. Undoubtedly, a clearer picture of the factors important for achieving good safety performance will emerge as more research evidence accumulates.

Another reason that these research findings are noteworthy is that most of the variables which appear to be playing a significant role in achieving a good mine safety record are within management's ability to control. The next section of this report lists some of the recommendations that the authors of these studies have made concerning strategies for improving mine safety. One should bear in mind that these recommendations are based on research findings that are subject to the above mentioned limitations.

RECOMMENDATIONS FOR IMPROVING MINE SAFETY

Many of the articles and reports cited in this review list specific recommendations for mine operators and other mining officials concerning the achievement of a good safety record. These recommendations mostly fall into one of the following nine categories: safety programs and their directors, labor-management interaction, industry commitment, top management commitment, training, employee motivation and/or incentives, supervisor–employee interactions, management planning, and absenteeism. The recommendations made concerning each of these topics are listed below. They vary widely in terms of the specificity of recommended action.

1. Safety programs and their directors

DeMichiei (9):

- Management must establish a formal safety and health program, effectively communicate that program to employees and seek labor's active participation in a joint implementation of the program.
- Management should commit the funds and peoplepower necessary to establish a safety department and give its personnel the authority to implement the safety and health program effectively.

Pfeifer (25):

- Safety directors' jobs should be redesigned to provide for more time in developing company safety programs.
- Safety directors should be given more authority in the area of safety.

2. Labor-management interaction

DeMichiei (9):

- Management and labor must establish open lines of communication so that problems affecting health and safety can be freely discussed and mutually resolved. Open lines of communication must exist between all levels of management and labor so that unsafe conditions or practices can be corrected and employees can feel free to discuss safety issues without fear of adverse action.
- Management and labor must both recognize that safety and health is a joint responsibility, which will only be achieved through cooperative efforts. Management and labor representatives must take the lead in cultivating a cooperative atmosphere at the mine via joint informational meetings, safety inspections of workplaces, and increased interaction with the general work force.
- Management should actively involve representatives of labor on issues concerning safety, health, and production.
- Labor representatives must support mine management when it is necessary to reprimand miners for unsafe acts.
- Labor representatives must instill a sense of responsibility and accountability in the work force for their actions and resultant impact those actions may have on fellow workers.
- Management should solicit the assistance of labor in identifying and correcting unsafe conditions and practices. Conversely, labor should accept the joint responsibility in this endeavor.

- Management should periodically review and solicit the opinions of miners concerning established mine policies and procedures to determine their effect on the miners' morale.

National Academy of Sciences (*21*):

- There is a need to establish joint labor-management safety committees at each mine.

Pfeifer (*25*):

- Miners need to be able to better communicate to management problems affecting their health and safety.
- Miners should be given a hand in the establishment of new company safety procedures, or, at a minimum, the reasons for new safety procedures should be better explained to miners.

3. Industry commitment

National Academy of Sciences (*21*):

- Encourage industry leaders to reinforce the value and importance of safety.

   ... saving lives and minimizing injuries are deeply held social values. However, it may be necessary for industry leaders to remind their peers that safety is as important a value to uphold as is producing coal at the lowest cost per ton.

- Encourage publication of annual rankings of companies by their injury rates.
- Publicize the evidence that productivity and safety can be positively related.

   The common denominator in achieving both a good safety record and productivity is competent management. A management that can plan well to increase production can also plan well to improve safety. Moreover, a management that shows concern about safety signals to its employees that it is concerned about their well-being, and, thus, deserves their contributions of skill and energy in improving productivity. Finally, a management that is willing to listen to employees' ideas for improving safety (which the researchers found associated with effective programs) is also likely to listen to employees' ideas for improving productivity.

4. Top management commitment

Pfeifer (*25*):

- Coal companies need to formalize safety as an organizational goal.
- Coal companies should communicate to workers both verbally and through

the behavior of management the importance of safety as an organizational goal.
- Coal companies need to place more emphasis on keeping good safety records and upholding the company safety record.

5. Training

DeMichiei (9):

- When formulating new training programs or revising existing programs, greater input should be solicited from labor by mine management. A thorough review by mine management, labor, and MSHA of existing training programs will ensure that such plans are tailored to individual mine needs.
- First line supervisors should be provided instruction and sufficient time for administering task training to subordinates.
- The means to measure the effects of both classroom and on-the-job training should be incorporated into training programs.

National Academy of Sciences (21):

- A major upgrading of the educational and training requirements that new and experienced miners, and also supervisors, must meet is needed.

Pfeifer (25):

- In addition to better training for miners, better training is needed for mine management beginning at the level of supervisor. Supervisors must be made aware that they serve as models for the miners working under them, and, consequently, they must follow all safety procedures.

Sanders (27):

- Training mine management in basic supervisory employee relations and planning skills would significantly impact on the injury rates of the mines.

6. Employee motivation and/or incentives

Pfeifer (25):

- Miners' jobs need to be redesigned in order to provide for satisfaction of miners' intrinsic needs such as recognition, responsibility, and variety of job tasks, and, so that they do not have to do many things at the same time, they do not have to work without clear-cut duties, and they do not have to answer to more than one person.
- Programs providing reinforcement for a variety of identifiable safe job behaviors need to be established for both underground miners and supervisors. Techniques, which should be used in providing incentives for safe job behavior, include making safety an integral part of worker performance evaluation and publicizing outstanding safety performance.

- Care must be taken in employing competition among workcrews to be sure that safety is included in the criteria of good job performance.

Goodman (17):

Goodman presents a great deal of advice regarding the appropriate design of bonus plans for improving underground coal mine productivity and safety.

7. Supervisor-employee interactions

Uslan (30):

- Supervisors can affect employee motivation to perform their jobs safely by: (1) encouraging employees to buy into explicit safety performance goals that are consistent with organizational objectives, (2) arranging conditions so that employees can accomplish goals, (3) determining rewards that employees desire and making these rewards contingent on high levels of safety performance, and (4) ensuring that employees understand the relationship between safety performance and the receipt of rewards.

DeMichiei (9):

- Management should periodically monitor on-the-job work procedures to ensure that labor has the necessary experience or qualifications to recognize adverse conditions.

8. Management planning

DeMichiei (9):

- Management should formulate, implement, and enforce systematic mining cycles and standardized work procedures.
- Management should develop a comprehensive approach for mine development that includes activities such as materials handling, transportation and installation of equipment parts, etc. "Many times such activities are not considered as an integral part of the production."
- Management, labor, and MSHA must ensure that mine plans incorporate measures necessary to adequately control the physical environment. "Too often management continues to implement minimum plan requirements when, in fact, additional measures are necessary."

Pfeifer (25):

- Coal companies should establish better programs for maintenance of equipment.

9. Absenteeism

DeMichiei (9):

- Management should establish and implement an absentee policy that is firm, but fair, taking into consideration extenuating circumstances that could adversely affect mine personnel.

Goodman (*18*)—strategies for *coping* with miner absenteeism:

- Coal mine operators could better cope with absent members of underground coal mining crews by increasing familiarity among replacements for absent miners. This can be accomplished by organizing pools of replacement workers, and by giving special on-the-job training to a replacement and to the adjacent worker before work begins, in order to help familiarize each miner with the practices of its partner. Pools of replacement workers could be organized by job categories. For example, certain replacements would work as miner helpers, others as car operators. The pool could be further organized by mine sections; i.e., when possible, certain workers would always be assigned to certain work areas with which they are relatively familiar.

Goodman (*15*)—strategies for *reducing* miner absenteeism:

- Improve hiring practices through use of better selection procedures and realistic job previews.
- Institute employee assistance programs, health education programs, and selected types of in-house medical services.
- Maintain a record of each employees' daily attendance and post it somewhere for employees to see.
- Train supervisors in what they can do to prevent chronic absenteeism.
- Use positive reinforcement programs.
- Ensure consistency in the use of progressive disciplinary procedures.

REFERENCES

1. Althouse, R., and J. Hurrell. An Analysis of Job Stress in Coal Mining. NIOSH (DHEW) 77-217, 1977, 145 pp.; NTIS PB 274-796.
2. Andlauer, P., and B. Metz. Le Travail en Equipes Alternates (Work By Rotating Teams). Physiologie du Travail-Ergonomie (Physiology of Work-Ergonomics), ed. by J. Sherrer, Masson (Paris), 1967, pp. 272–281.
3. Bell, C. The Hecla Story: Organization Development in the Hard-Rock Mining Industry. Paper in Human Engineering and Human Resources Management in Mining. Proceedings: Bureau of Mines Technology Transfer Seminar, Pittsburgh, PA, July 7–8, 1987; St. Louis, MO, July 15–16, 1987; and San Francisco, CA, July 21–22, 1987, comp. by staff. BuMines IC 9145, 1987, pp. 138–148.
4. Bennett, J., and D. Passmore. Probability of Death, Disability, and Restricted Work Activity in United States Underground Bituminous Coal Mines, 1975–1981. J. Saf. Res., v. 15, No. 2, 1984, pp. 69–76.

5. Braithwaite, J. To Punish or Persuade: The Enforcement of Coal Mine Safety. State Univ. NY Press (Albany), 1985, 206 pp.
6. Buller, P., and C. Bell. Effects of Team Building and Goal Setting on Productivity: A Field Experiment. Acad. Manage. J., v. 29, No. 2, 1986, pp. 305-328.
7. Cohen, A. Factors in Successful Occupational Safety Programs. J. Saf. Res., v. 9, No. 4, 1977, pp. 168-178.
8. DeJoy, D. Attributional Processes and Hazard Control Management in Industry. J. Saf. Res., v. 16, No. 2, 1985, pp. 61-71.
9. DeMichiei, J., J. Langton, K. Bullock, and T. Wiles. Factors Associated With Disabling Injuries in Underground Coal Mines. MSHA, June 1982, 72 pp.
10. Edwards, S. Quality Circles Are Safety Circles. Natl. Saf. News, June 1983, pp. 31-35.
11. Fiedler, F. Structured Management Training in Underground Mining—Five Years Later. Paper in Human Engineering and Human Resources Management in Mining. Proceedings: Bureau of Mines Technology Transfer Seminar, Pittsburgh, PA, July 7-8, 1987; St. Louis, MO, July 15-16, 1987; and San Francisco, CA, July 21-22, 1987, comp. by staff. BuMines IC 9145, 1987, pp. 149-153.
12. Fiedler, F. E., C. H. Bell, Jr., M. H. Chemers, and D. Patrick. The Effectiveness of Organization and Management Training on Safety and Productivity in Metal/Non-Metal Underground Mining (contract J0387230, Perceptronics, Inc.). BuMines OFR 191-84, 1983, 296 pp.; NTIS PB 85-163285.
13. Filley, A., R. House, and S. Kerr. Managerial Process and Organizational Behavior. Scott Foresman, 1976, 558 pp.
14. Gaertner, G. H., P. D. Newman, M. S. Perry, G. P. Fisher, and K. Whitehead. Determining the Effects of Management Practices on Coal Miners' Safety (contract JO145029, Westat, Inc.). BuMines OFR 39-88, 1987, 348 pp.; NTIS PB 88-221445.
15. Goodman, P. S. Analysis of Miners' Job Attendance Behavior and Its Relationship to Miners' Accidents and Injuries—Final Report (contract JO328033, Carnegie-Mellon Univ.). BuMines OFR 60-86, 1985, 278 pp.; NTIS PB 86-216306.
16. ———. Assessing Organizational Change: The Rushton Quality of Work Experiment. Wiley, 1979, 391 pp.
17. Goodman, P. S. Determining the Effect of Incentive Programs on the Occurrence of Accidents, Injuries, and Productivity—Final Report (contract JO145012, Carnegie-Millon Univ.). BuMines OFR 47-88, 1987, 143 pp.; NTIS BP 88-234257.
18. Goodman, P., and S. Garber. Absenteeism and Accidents in a Dangerous Environment: Empirical Analysis of Underground Coal Mines. J. Appl. Psych., v. 73, No. 1, 1988, pp. 81-86.

19. Miles, W. Incentive Programs Get Varied Response From Management. Natl. Saf. Counc. Min. Newslett., Mar.-Apr., 1986, pp. 1–3.
20. Mills, T. Altering the Social Structure in Coal Mining: A Case Study. Mon. Labor Rev., v. 99, No. 10, 1976, pp. 3–10.
21. National Academy of Sciences—National Research Council. Toward Safer Underground Coal Mines. NAS, 1982, 190 pp.
22. Page, S., J. Volkwein, and F. Kissell. Some Continuous Sections Can Cut More Than 1,000 Tons Per Unit. Coal Age, Jan., 1987, pp. 51–55.
23. Peters, R. Foreign Literature on Environmental and Personal Factors Affecting the Safety and Productivity of Miners: A Topical Listing and Annotated Bibliography of Recent Research. BuMines OFR 198-83, 1983, 91 pp.; NTIS PB 84-127687.
24. Peters, R., and L. Schaffer. Field Tests of a Model Health and Safety Program for the Mining Industry. BuMines IC 9075, 1986, 36 pp.
25. Pfeifer, C., J. Stefanski, and C. Grether. Psychological, Behavioral, and Organizational Factors Affecting Coal Miner Safety and Health (DHEW contract HSM 99-72-151). DHEW, 1976, 319 pp.; NTIS PB 275 599.
26. Rhoton, W. A Procedure To Improve Compliance With Coal Mine Safety Regulations. J. Organizational Behav. Manage., v. 2, No. 4, 1980, pp. 243–249.
27. Sanders, M., T. Patterson, and J. Peay. The Effect of Organizational Climate and Policy on Coal Mine Safety (contract HO242039, U.S. Dep. Navy). BuMines OFR 108-77, 1976, 180 pp.; NTIS PB 267 781.
28. Schaffer, L., and D. Atchison. Research To Improve Health and Safety Programs in the Mining Industry. Volume II (contract HO308076, Woodward Assoc., Inc.). BuMines OFR 6-85, 1983, 197 pp.; NTIS PB 85-151017.
29. Trist, E., G. Susman, and G. Brown. An Experiment in Autonomous Working in an American Underground Coal Mine. Hum. Relat., v. 30, No. 3, 1977, pp. 201–236.
30. Uslan, S., H. Adelman, and R. Keller. Testing the Effects of Applied Behavioral Analysis and Applied Behavioral Management Techniques on the Safe Behaviors of Salt Mine Personnel (contract JO166137, Salt Inst.). BuMines OFR 44-80, 1978, 44 pp.; NTIS PB 80-171309.
31. Wagner, J. Time-Of-Day Variations in the Severity of Injuries Suffered By Mine Shiftworkers. Paper in Proceedings of the 32nd Annual Meeting of the Human Factors Society. Anaheim, CA, 1988, 11 pp.; available upon request from J. Wagner, BuMines, Minneapolis, MN.
32. Webber, R. Management: Basic Elements of Managing Organizations. Irwin, 1975, 175 pp.

Appendix *E*

Major Incidents

In the National Safety Council's publication, *Shiftwork Safety and Performance,* by Glenn McBride and Peggy Westfall, an excellent appendix is provided of major incidents and costs, which we have reprinted here with permission from the National Safety Council.

This table of sample incidents and costs was generated from accessible CNN, AP, safety, and press reports and is incomplete. We do not claim to maintain the absolutely latest data and would appreciate updates from industry. The costs are estimated by journalists, court reports, and the businesses involved.

Incidents and Costs

Incident	Type Incident	Details	Costs	Fines/Penalties Reported/Proposed	Date
Texas City Explosion	French ship Grandcamp exploded while docked.	Ammonia nitrate blew up and next day more fertilizer on the High Flyer blew up	•576 fatalities •5,000 injured		4/16/47
New York Emergency Room Incident	•Hospital in New York. A milestone legal case.	•Two doctors who treated patient made mistakes. Had worked 18 hours straight.	•Regulations to limit on-duty hours which State expects $3.1 billion over next 10 years		

385

Incident	Type Incident	Details	Costs	Fines/Penalties Reported/ Proposed	Date
Soviet Nuclear Incident at Chelyabinsk-65 near Kyshtym in Urals.	•Apparently, tank of radioactive waste exploded. Discharged 20 million Curies of radiation.	•Probably contaminated 357 square miles 10,000 evacuated.	•Possibly several hundred fatalities •200 million rubles.		9/29/57
Farmer's Export Grain Elevator Explosion		Grain dust was ignited by a spark.	•18 fatalities		12/27/77
Three-Mile Island, Middletown, Penn.	Nuclear reactor partial meltdown	Equipment error and human error.			3/28/79
Galveston Bay Oil Spill	Tanker spilled 10,700,000 gallons of oil.	Two ships had collided in the Bay			11/1/79
Polyethylene Plant Explosion	Dow Chemical, Texas suffered major explosion.	The explosion occurred during a shutdown	•6 fatalities.		10/13/81
Mexico City Gas Storage Explosion	Four spherical 420,000 gal. tanks were ignited from a propane truck at loading dock.	Homes were allowed to be built near the facility	•30 acres of homes destroyed and another 30 acres damaged •540 fatalities •2,200+ injured •10,000+ homeless.		11/19/83
Houston Nursing Home Incident	84 year-old woman was strangled by a restraint device.	Human error in a medical setting—extended over three nights.		$40,000,000 awarded to family of the woman	1984
Vila Soco Pipeline Fire	Pipeline gasoline blaze exploded and burned at over 1000 degrees Celsius through a Brazilian village.	The wrong pipeline was opened the day preceding the fire.	•500+ fatalities (Child casualties under age 5 had to be estimated since they were totally incinerated.)	•Petrobras paid hospital costs and damages.	2/25/84

Major Incidents

Incident	Type Incident	Details	Costs	Fines/Penalties Reported/Proposed	Date
Louisiana Coast Oil Spill	Tanker Alvenus spilled 2,800,000 gallons of oil	The ship struck bottom and ruptured off Louisiana.	•Galveston beaches and Port Arthur coastline was damaged	•Foreign owner paid $1,100,000 to State of Texas to settle the lawsuit	7/30/84
British Tanker Spill in Louisiana	Tanker spilled oil on Louisiana coast.	Tanker ran aground and spilled 2,800,000 gallons of crude, causing 85-mile slick.	•Slick drifted into Galveston Island and Port Arthur shores		7/30/84
San Juanico Pemex Gas Explosion	A series of liquified gas storage explosions in San Juanico, Mexico.	Fireball flashed through suburban area at 5:43 a.m.	•503 fatalities •4,000+ injured.	Pemex was held liable by the federal attorney general. •By 1986, Pemex had paid $5,000,000 in claims.	11/19/84
Mont Belvieu Plant Explosion	Warren Petroleum plant exploded.	150 workers reached safety minutes before the ignition!	•2 fatalities		11/05/85
Kerr-McGee, Gore, Oklahoma Nuclear Mishap	Cylinder of nuclear material burst.	Cylinder was being improperly heated	•1 fatality •100 hospitalized.		1/06/86
Challenger Space Incident	Shuttle failed due to a faulty booster rocket.	Sleep loss and overtime were cited.	•Lives of all aboard. •Significantly impacted progress of NASA programs	•Morton Thiokol changed name to Morton International, Inc due to publicity. •Invested $600,000,000 in air bags for automobiles to improve image	1/28/86

Incident	Type Incident	Details	Costs	Fines/Penalties Reported/Proposed	Date
Chernobyl	Nuclear plant meltdown and radiation release •Poor engineering and operation combined.	Released 50 million Curies of radiation into surrounding area.	•250+ deaths •$26 billion planned to move 200,000 additional residents •$2 billion planned to rebury the plant.		4/26/86
Ohio Tank Car Gas Release	CSX train derailed leading to explosion and burning chemicals	Tank car of white phosphorus released gas over Dayton suburb.	•25,000 evacuated	•Governor called for more state regulations.	7/8/86
Texas Refinery Acid Release	Hydrofluoric acid release from Marathon Petroleum refinery led to evacuation.		•225 injured •4,000 residents evacuated		10/30/87
Monongahela River Spill	Ashland oil tank ruptured and spilled 3,800,000 gallons of diesel fuel. 860,000 escaped into river.	•70+ communities lost water supply			1/2/88
Thompsontown, Penn. Train Collision	Conrail suffered head-on collision of two freight trains.	Engineer and crew possibly sleeping or experiencing micro-sleep.	4 lives and $6,000,000		1988
Alaskan Oil Spill	987-foot Tanker smashed into Bligh Reef and spilled 11,000,000 gallons of oil into Prince William Sound.	•Capt. Joseph Hazelwood left the bridge during maneuvers. He and his crew have been blamed by gov't officials and others •Exxon later felt that a Sperry control system might have caused the incident.	This incident was eventually labeled as a HUMAN FATIGUE incident by the NTSB investigators.	•On-going. •U.S. Congress passed a bill allowing states to adopt stricter spill liability laws than the federal govt. requires.	3/24/89

Major Incidents

Incident	Type Incident	Details	Costs	Fines/Penalties Reported/ Proposed	Date
Houston Ship Channel Barge and Tanker Collision	252,000 gallons of slurry oil were spilled into the Ship Channel.				6/23/89
Delaware River Oil Spill	Uruguayan Tanker spilled 300,000 gallons of fuel oil	Tanker ran aground in the Delaware River.			6/24/89
Phillips Petroleum Pasadena Explosion	Gas release led to explosion which destroyed portion of a polyethylene plant	Phillips Petroleum's President Glenn Cox said the company's own investigation showed the explosion "was the result of a departure from established routine procedures. . ."	•23 fatalities •314 injuries •Phillips experienced $434 million decrease in net income that year	OSHA first proposed $6,400,000 in fines. Later reduced them in exchange for promise to institute process safety management procedures at Pasadena and three other plants.	10/23/89
Arthur Kill Oil Spill	Exxon refinery pipeline spilled 567,000 gallons of heating oil.	The underwater pipeline running from an Exxon refinery ruptured and spilled the oil into Arthur Kill.			1/1/90
Southern California Oil Spill	American Trader (charted by British Petroleum) spilled 400,000 gallons of Alaskan crude oil	The ship was "gored" by its own anchor!			2/7/90
Mega Borg Tanker Spill	Tanker spilled 38,000,000 gallons of light crude into the Gulf of Mexico, off Galveston Island.	Explosion occurred aboard the supertanker.			June, 1990

Incident	Type Incident	Details	Costs	Fines/Penalties Reported/ Proposed	Date
Shinoussa Tanker Spill	Tanker spilled 700,000 gallons of crude oil into Galveston Bay.	Tanker and barge collided in the Houston Ship Channel			July, 1990
Cincinnati Resins Plant Explosion	The BASF plant suffered an explosion and fire.	•Workers had a few seconds warning. •A flammable solvent venting into the plant may have been the cause.	•1 fatality. •56 injuries.	•$1,100,000 in OSHA fines.	July, 1990
Channelview, Texas Chemical Plant Explosion	Houston Arco Channelview plant suffered explosion that burned city block-sized area. The fire lasted over four hours.	Inadequate training and excessive overtime work have been mentioned as possible causes of the accident.	•17 fatalities	•$3,480,000 in fines.	7/5/90
Philadelphia Electric Company Fines	Nuclear Plant error was reported to NRC	Operators sleeping in control centers		•$1,250,000 NRC fine. •Also: $500 to $1,000 fines to 33 workers and supervisors	1991
Staten Island Ferry terminal Fire		Unauthorized people (smoking crack?) in the building probably set off fire in an 8 foot-high cockloft.	•$50,000,000 estimated to rebuild the terminal building.		1991
Lake Charles Refinery Explosion	Citgo Petroleum's Louisiana refinery suffered a major explosion and fire.		•6 fatalities •6 injuries	•$5,800,000 in OSHA fines. •Promise to enhance safety at Lake Charles and two other locations.	3/3/91

Major Incidents

Incident	Type Incident	Details	Costs	Fines/Penalties Reported/ Proposed	Date
Nitroparaffins Plant Explosion	IMC Fertilizer operated the plant for Angus Chemical.		•8 fatalities •42 worker injuries •70 resident injuries	•$9,800,000 IMC Fertilizer Inc. OSHA fine. •$200,000 Angus Chemical OSHA fine. •Both companies also promised to institute or upgrade process safety management procedures at four IMCF Angus and IMCF facilities. The analyses must cover human error and startup-shutdown conditions.	5/01/91
Chicago Tunnel Flood	100 year-old freight tunnel ruptured and spilled millions of gallons of water into downtown Chicago.	•Although he saw the damaged wall, an engineer did not react and "waited for photos from the drugstore." •Inspectors looked at only one bridge because parking was difficult.		•On-going. •Mayor Daley fired at least two officials and suspended others who apparently knew about the damage.	4/13/92

Index

Aaron, H., 7
Abilene Paradox, 16–18
Accident proneness, 32, 135, 213–219, 310
Accountability, 3, 32, 256
Alcohol, 63, 117, 136, 153, 258, 324
Alertness, 129–130
Alkov, R., 157
American Society of Safety Engineers (ASSE), 305
Anxiety, 153
Argyris, C., 113, 169–170, 267
Arrangement principle, 86
Arousal state, 174–176, 268
Assessment criteria, 56, 73
Association of American Railroads (AAR), 75, 248
Attitude
 components of, 174
 development, 207, 266, 268, 312
 formation, 175
Attitudes, 206–207
Attitudinal state, 174, 263–266
 conditions, 263
 consequences, 264–265
Aversives, 208–209, 265

Bailey, C., 79
Barrier removal, 269, 287
Behavioral change, 139
Behavior modification, 139, 260, 296
Better Sleep Council, 152–153

Bhopal (India), 10, 12, 68–70, 151–152, 234
Biorhythms, 176–177, 266–267
Blanchard, K., 238
Blunder Book, The, 5, 7
Bond, N., 13
Boredom, 149, *151,* 267
Boss evaluation, 269, 313–315
Bracey, H., 307–308
Brewer, J., 63–64
Broadbent, D., 147
Brody, L., 218
Brown, R., 248

Capacity, 31, 125. *See also* Subsystems, human
Catastrophe(s)
 attributes for, 68–69
 causes of, 67–73
 control, 12
 predictability of, 73
Catastrophe vs. attribute matrix, *70*
Challenger (space shuttle), 10, 12, 68–70, 233–234
Change, organizational, 278
Change agent, 310
Chapanis, A., 4, 58
Chemical Manufacturers Association (CMA), 18
Chernobyl, 10, 12, 68–71, 151
Clemente, R., 7

393

Climate, 65, 164–168. *See also* Culture
  analyzing, 168–169, 269, 282–287, 311
  variables, 65
Coaching, 139
Cobb, T., 7
Coding principle, 90
Columbus, C., 5
Common sense, 227–228
Compatibility principle, 86
Conflict theory, 113, 169, 190
Conover, D., 98, 131, *132*
Cooper, K., 25–29
Contracts, 300
Coplen, M., 152
Cost control, 32
Crisis intervention, 260, 263, 296
Criteria for safety excellence, 56
Culture, 32, 34, 65–68. *See also* Climate
  and safety, 66
  assessing, 73–79
  -caused error, reducing, 96–97
  causes, 63–64
  definition, 65
Cumulative trauma disorders, 81, 91–93, 105–106
  analysis, 335–338

DC –10 (aircraft), 10
Dalton, G., 279
Deatheridge, B., 129
Decision to err, 31, 32, 195
  attitudes, 206–209
  peer group, 199–206
  reducing, 247–251
  reinforcement, 196–198
Design, as cause of error, 81, 85. *See also* Ergonomics, Human factors engineering
  reducing, 97–98
Design errors, 10
Diekemper, R., 42–43
DiMaggio, J., 7
Discipline, 185
Dissociability principle, 89
Drugs, 63, 117, 136, 258, 324
Dukes-Dubos, F., 82

Eastman Kodak, 143
Edsel, 7
Emotional factors, 27

Employee
  behavior modification, 139–142
  interviews, 75, 137–138
  reference checks, 138
  selection and placement, 136–137
  surveys, 75
  training, 139
Employee Assistance Programs (EAPs), 117
Environmental load, 147–148. *See also* Work environment
Ergonomics, 82, 92–93
  CTD analysis, 105–106, 335–338
  data, 339–353
  systems analysis, 99–100, *101*
  task analysis, 100, *102*, 105
Error. *See* Human error.
Error-provocative situation, 57. *See also* Human error
Euninger, M., 242, 251
Executive liability, 11
Expectations, 90, 247
Exxon Valdez, 10, 151–152

Fatigue, 136, 149, *151,* 267
Ferrell, R., 29
Fight or flight response, 114
Fletcher, J., 51
Force field analysis, 237–238, 269–270
Forced entry principle, 89
Foxx, J., 7

Geller, E.S., 177
Gellerman, S., 282–286
Goal setting, 183
Goldberg, M., 5, 7
Graham, S., 111–112
Grose, V., 232–237
Group
  analysis, 319
  approaches, 300–302
  effectiveness, 201–203
  norms, 200

Hackman, J., 198, 199
Handy, C., *155, 204,* 273
Hannaford, E., *175, 176*
Harvey, J. 16–18
Healthy reactions, 116
Hearing, 127
Heinrich, H., xiii-xvii, 56

Index

Hersey, P., 238
Herzberg, F., 113, 169
Holmes, T., 157, 159
Hostility, 113
Human Development Corporation, 248
Human error
  aberrations, due to, 15
  causation model, 29–31
  causes, 14, 31–34, 63–64
  costs, 11
  definitions, 3–4
  influences, 25–29
  outcomes, 12–14
  rates, 15
  reducing, 95–106, 255–271
  spectrum, 12
  types, 14, 15–16
Human factors engineering, 81–82. *See also* Ergonomics
Human organization, The, 65, 75
Human performance, improvement model, 57
Human subsystems. *See* Subsystems, human
Hypochondria, 120

ILO/WHO Committee on Occupational Health, 117–119
Information processing subsystem, 89, 125, 267
In Search of Excellence, 65
Intentional errors, 19. *See also* Human error.
Intelligence, 130, 133
Interviews, 75
Invariance principle, 89
Inverse performance appraisals, 270, 287

Jerison, H., 147
Job safety analyses (JSAs), 66
Job Demands and Worker Health, 132, *133*
Job stressors, 179
Johnson, W., 7
Johnson, W.G., 57–58, 97
Justice Department, 11

Kamp, J., 119–122
Katz, D., 230–232
Kennedy, J.F., 18
King, P., 187–190

Kinkade, R., 99, *126*
Kletz, T., 15–16
Knowledge (of job), 136. *See also* Training.
Koufax, S., 7
Kreitener, R., *141*

Ladou, J., 184
Larson, J., 130
Lawler, E., *198,* 294, 295
Leveling analysis, 270, 287, 290
Levens, E., 308–310
Levinson, H., 274–278
Lewin, K., 237–238
Life Change Units (LCUs), 157, 267–268
Likert, R., 65, 75, 167–168, 169, 170–171
Lippitt, G., *87, 280*
Load, 31, 145. *See also* Stress
  long-term, 157–160
  short-term, 145–156
Logic diagrams, 59, *60, 61*
Logical decision to err. *See* Decision to err
Long-term load. *See* Load
Lowe, G., 152
Lower management. *See also* Management
  roles, 290–294
  tasks, 294–302
Luck, 34
Luthans, F., *141, 142*

Mager, R., 207–208, 263–265
Management
  -caused error, reducing, 95–96
  lower, 290–302
  roles, 273–278, 290–294
  safety procedures, 256–257
  safety systems, 42–56
  styles, 209–210
  system errors, 10
  systems safety outline, 37–42
  upper, 273–290
Management oversight and risk tree (MORT), 37, *38,* 57–58, 97
Managing Employee Stress, 73–79
Mason, C., 223–226
Matthyson, B., 51, 53–54
Maturity, 190
McBride, G., 150–153, 385
McFarland, R., 14, 82, 84, 85–86
McGrath, J., *147,* 153–154, *156, 157*
McKelvey, R., 248

McKenna, R, *140*
Mechanical ability, 135
Mental condition, 32. *See also* Accident proneness
Minor, J., 63–64
Mitler, M., 151–152
Modeling (behavior), 266. *See also* Attitudinal state
Moore-Ede, M., 111
Motion sensors, 129
Motivation factors, 163, 171
Motivational state, 163
Motor ability, 27
Myers, M., 166–167

National Commission on Sleep Disorders, 112, 151
National Institute for Occupational Safety and Health (NIOSH), 132, 134
National Occupational Safety Association (NOSA), 51, 53–54, 55
National Safety Council, 51
National Safety Management Society (NSMS), 29
Noise, 26. *See also* Stress
Nontraditional controls, 260–271
Nuberg, A., 12

Occupational Health and Safety Act (OSHA), 11, 56, 66, 91, 105
Older workers, 28
Occidental Petroleum, 71
Opportunities for human error, *20*
Osuna, J., 59, *61*
Outcomes of error, 12–14
Overload, 31, 109–112
  information-processing, 267
  psychosocial factors, 117–122
  reactions to, 112–117
  reducing, 182–192
Overtime, 109–111

Parsimony principle, 89
Participation, 186–190
Peer group, 199–201
  pressure, 16–18, 32, 173–174, 270
Perceived probability, 32
Perception, 27
  and common sense, 227–228
  of culture, 32. *See also* Climate, Culture
  of peer thought, 16–18

of risk, 223–227
of safety, 228–230
surveys, 75, *78, 79,* 96, 247
Performance shaping factor (PSF), 21–23
Personality, 27, 133–135, 172–173
Peters, G., 3–4
Peters, R., 355–383
Peters, T., 65
Physical ability, 16, 26
Pinto (gas tank), 10
Piper Alpha disaster, 70–72
Placement, employee, 136–138
Porter, L., 294, *295*
Positive reinforcement, 247–251, 265–266
Predictability of error, 73
Priorities, 209, 270, 315, *316, 317,* 319
Profiling, 51–56
Proneness. *See* Accident proneness
Psychological
  and personality traits, 133–136
  characteristics, 27
Psychosocial factors, 117–122

Rahe, R., 159
Random errors, 20. *See also* Human error
Random variability, 20
Recht, J.L., 14–15
Reference checks, 138
Reinforcement, 196–198, 247–251
Results of error, 10–12
Reitsesk, J., *140*
Risk management, 67–73
Risk taking, 130
Robertson, S., 110–111
Role definition, 184–185
Ruth, B., 7

Safety climate survey, 78
Safety professional
  role, 305–310
Safety program review, 37, *38*
Safety programs, essentials of, 54, 66
Safety systems
  management, 42–56
Sanheon, J., 177
Sayles, L., *151*
Scanlon, B., 165–166, 168
Schugsta, P., 215, *216, 217, 218*
Schulzinger, M., 214–215
Selection, employee, 136–138

Index

Sensory motor tests, 130
Sichel, H., 135
Shaw, L., 135
Sherif, M., 203
Shift work, 111–112, 150–153
Shortcuts, 19
Short-term load. *See* Load
Sinaiko, H., 126, *127*
Simplicity principle, 86
Situationality principle, 88
Skill, 136. *See also* Training
Skinner, B.F., 141, 196
Skin sensors, 129
Smell, 129
Social Research, Inc., 164–165
Spartz, D., 42–43
Spreigel, W., 137
Staff safety specialist. *See* Safety professional
State, 31
  arousal, 174–176
  attitudinal, 174
  biorhythmic, 176–177
  motivational, 163–174
Stereotypes (design), 90–91
Strauss, G., *151*
Stress, *23,* 26, 73–79, 136, 153–157. *See also* Load
  tests, 323–334
Subsystems, human, 125–131
  considerations, 131–136
Subsystems, worker's
  information-processing, 89–90
  responding, 90–91
  sensing, 86–89
Supervisory performance model, 256, *257*
Surveys, 75
System-caused human error, 54–56
  reducing, 95–106
Systems analysis, 99–100
Systems failure, 29, *30,* 31, 37
Systems safety, 259
  outline, 37–42
Systematic errors, 20–21
Systematic variability, 20

Task analysis, 100, *102,* 103–105
Taste, 129
Temperature, 27–28
Temporary capacity reducers, 136
Three Mile Island, 10, 12, 16, 67–71, 151, 234
Thermal stress, 26
Toch, H., 227–228
Traditional controls, 257–259
Training, 16, 65, 136, 185, 296
Traps, 32
Truman, H., 7
Tye, J., 51

Unintentional errors, 19. *See also* Human error
Upper management. *See also* Management
  changes, 278–290
  roles, 273–278

Van Cott, H., 99, *126*
Vespucci, A., 5
Vilardo, F., 88–89
Vision, 126–127
Volard, S.V., 25–29

Watergate, 18
Waterman, R., 65
Webster, J., 138
Westfall, P., 150–153, 385
Wood, F., 86, *87*
Woodson, W., 97, 98, 131, *132*
Work environment, 26
Work load, 31. *See also* Load, Overload
Work scheduling, 183–184
Worker role, 302
Worker-safety analysis, 271
Worker subsystems. *See* Subsystems, worker's
Worker's compensation, 11–12
Workers vs. machines, 131–132

Yariger, A., *154*
Young, C., 7

Zebroski, E., 67–73